PETER POND

Fur Trader and Adventurer

by
Harold A. Innis

© 2011 Benediction Classics, Oxford.

Contents

PREFACE	1
I. EARLY LIFE AND THE ARMY	3
II. THE APPRENTICE TRADER	12
III. THE MISSISSIPPI	15
IV. THE NORTHWEST COMPANY	38
V. THE MAN AND HIS WORK	61
BIBLIOGRAPHY	77

PREFACE

It has been a common assumption among writers on the fur trade that the more important traders and explorers were born in Scotland but in the formative period of the trade immediately following the Conquest and previous to the formation of the Northwest Company, traders born in the Colonies occupied an important place. Two of these traders, Alexander Henry, born in New Jersey in Aug. 1739, and Peter Pond, born in Milford in the county of New Haven in Connecticut on Jan. 18, 1740, exercised a profound influence on exploration and trade in the Northwest. They were born within a year of each other and there were two interesting coincidences in later life. At the surrender of Fort de Levis in 1760 and on the journey down the St. Lawrence to the attack on Montreal, Peter Pond was a commissioned officer under General Amherst, and Alexander Henry was a merchant engaged in supplying the commissariat. In 1775 Alexander Henry proceeded from Michilimackinac to Grand Portage and to the Northwest on his first visit and on Aug. 18 of that year on Lake Winnipeg he was joined by Peter Pond who was also on his first visit to that country. The work of Alexander Henry has been generally known through the editions of his [Pg viii] *Travels and Adventures in Canada and the Indian territories* but the name of Peter Pond has been almost forgotten. This neglect has been due largely to the scarcity of material concerning him. His crowded life left little room for a formal education and his illiteracy was a serious handicap to the attainment of a permanent place in Canadian history. His journal was apparently written in the latter part of his life probably after he was sixty years of age and after he had ceased to have any connection with the fur trade. Apparently a large part of it has been destroyed but the remainder which has been printed in the Wisconsin Historical Collections and elsewhere is an extremely interesting and valuable record of the activities of his early life. From a study of this record it is possible to gain an insight into the later life of the man, a knowledge of which must be gleaned from scanty and biased information supplied by contemporaries.

My attention was attracted to the work of Peter Pond through a study of the early history of the Northwest Company. The organization of the technical side of the trade which followed the expansion to the Saskatchewan and especially to Athabaska and which made possible the [Pg ix] evolution of the Northwest Company was to a large extent a result of the efforts of Peter Pond. He was the first white man to cross the Methye Portage to the drainage basin of the Mackenzie river. To understand the history of the fur trade of the Northwest during the formative stages of the Northwest Company I found it necessary to attempt a biography of Peter Pond. His contributions to Canadian development, and the Northwest Company was a crucial organization, appeared to warrant a general appreciation of his position as one of the fathers of Confederation. The vilification of his enemies has up to the present made this estimate impossible.

I should like to thank Mr. L. J. Burpee for his kindness in arranging for the presentation of a part of this study before the Royal Society of Canada at Winnipeg, in May 1928, as well as for other kindnesses.

H. A. I.

I. EARLY LIFE AND THE ARMY

Peter Pond wrote regarding his ancestry "It is well known that from fifth gineration downward we ware all waryers ither by sea or land" and the available evidence in the Pond genealogy lends support to his statement. The date of arrival of the first Pond in New England has not been definitely established, but it is probable that he was one of two sons of William Pond, a neighbor of John Winthrop, of Groton, Suffolk who arrived with the latter at Salem, Massachusetts in the "great fleet" of 1630. Winthrop refers in one of his letters to John Pond and it is probable that the other son was Samuel and that he migrated from Salem with various Dorchester settlers to Windsor, Connecticut. In any case a Samuel Pond married Sarah Ware at Windsor on November 18, 1642, and a son, named Samuel Pond, was born on March 4, 1648. Six years later on March 14, 1654 the father died leaving his widow a small estate of 130 pounds. She apparently moved to Branford, Connecticut, and on July 6, 1655 married John Linsley at that place. The son became one of the charterers of Branford and in 1672 became a freeman at Hartford. He was deputy to the General Court for Branford in 1678, 1682-3 (seargeant of "ye trainband" in the latter year), and 1687, and was made lieutenant in 1695. On February 3, 1669 he was married to Miriam Blatchley of Hartford and on July 1, 1679, a son, Samuel Pond, the third, was born. On June 8, 1704 this son was married to Abigail Goodrich, of Branford, a daughter of Bartholomew Goodrich, who had been made lieutenant in 1695. A son, Peter Pond, was born at Branford on January 22, 1718. The latter married Mary Hubbard, a daughter of Zachariah Hubbard of Boston, probably in 1739, and Peter Pond, the second, was born in Milford on Jan. 18, 1740 the eldest of nine children. His mother died on June 16, 1761 at the age of thirty seven and his father in 1764.

PETER POND: FUR TRADER & ADVENTURER

Little is known of the early life of Peter Pond. According to the rolls of the Suffolk County Regiment he enlisted on April 17, 1759 as a shoemaker by trade and it is probable that his father was a shoemaker also. At the age of sixteen he writes that his parents forbade him to join the army "and no wonder as my father had a larg and young famerly I just began to be of sum youse to him in his afairs." The spelling of this journal warrants the conclusion that little time was available to acquire a formal education. The son of a shoemaker in a large family probably suffered from handicaps.

His account of his early life begins with enlistment in the colonial army. In 1755 Braddock was killed on the banks of Monongahela river with heavy losses to his troops. Pond writes "A part of the British troops which ascaped cam to Milford". Attracted by tales of adventures in the service he was determined to enlist. "Toward spring government began to rase troops for the insewing campaign aganst Crown Point under the comand of General Winsloe. Beaing then sixteen years of age I gave my parans to understand that I had a strong desire to be a solge. That I was detarmind to enlist under the Oficers that was going from Milford and joine the army.—the same inklanation and sperit that my ancesters profest run thero my vanes.—and indead so strong was the propensatey for the arme that I could not with stand its Temtations. One Eaveing in April the drums an instraments of Musick were all imployed to that degrea that they charmed me.—I found miney lads of my acquantans which seamd determined to go into the Sarvis. I talkt with Capt. Baldwin and ask him weather he would take me in his Companey as he was the recruiting offeser. He readealey agread and I set my hand to the orders."

Peter Pond was a recruit in the seventh company of the Connecticut regiment under Capt. David Baldwin. "My parans was so angry that thay forbid me making my apearance at home. I taread about the town among my fello solgers and thought that I had made a profitable Exchange giting a rigimintal coate—At length the time came to report. Early in June we imbarked on bord a vessel to join the arme at the randivoere. We sald from Milford to New York proceeded up North river and arrived safe at Albany. I cam on smartly as I had sum of my Bountey money with me. I did not want for ginger bread and small bear and sun forgot that I had left my parans who were exseedingley trubled in minde for my well-fair. After taring thare sum weakes the Prinsabel part of the Armey got togather and we proceaded up to the Halfmoon and thare lay til the hole of the Armey from differant parts of the hole countray got to gather. In the meantime parties and teamsters ware imploid in forwarding provishon from post to post and from Forte Eadward to the head of Lake George. It was supposed that we should crose Lake George and make a desent on ticondaroge but before that coud be a Complished the sumer ended. Fall of year Seat in and we went to work at the fort George which lay on the head of the Lake by that name. In November it Groed two cold to sleap in tents and the men began to Mutanie and say that thay had sarved thare times out for which thay ware inlisted and would return home after satisfying them with smooth words thay ware prevailed on to prolong the campaign a few weakes and at the time promest by the Gineral the camp broke up and the troops returned to thare respective plasis in all parts of ye country from which thay came. But not without leaving a grate number behind which died with the disentary and other diseases which camps are subject to appesaley (especially) among raw troops as the Amaracans ware at that time and they Beaing strangers to a holesome Mod of cookeraray it mad grate havock with them in making youse of salt provisions as they did which was in a grate part Broyling and drinking water with it to Exses".

This introduction to army life had satisfied him for the time and the following year did not find him anxious to enlist. "The year insewing which was 57 I taread at home with my parans so that I ascaped the misfortune of a number of my countrey men for Moncalm came against fort George and capterd it and as the Amaracans ware going of for fort Edward a Greabel to ye capatalasion (capitulation) the Indians fel apon them and mad grate Havack."

But in spite of the disasters of that campaign two years of inactivity proved unbearable. "In ye year 58 the safety of British Amaraca required that a large arme should be raised to act with the British Troops against Cannaday and under the command of Gineral Abercrombie against ticonderoge. I found tareing at home was too inactive a life for me therefore I joined many of my old Companyans a secont time for the arme of ye end of the Campain under the same offisers and same regiment under the command of Cornl Nathan Whiting"—of the second Connecticut regiment. "In the Spring we embarked to gine the arme at Albany whare we arrived safe at the time appointed. We ware emploid in forwarding Provishuns to Fort Edward for the youse of the Sarvis. When all was readey to cross Lake George the armey imbarked consisting of 18000 British and Provincals in about 1200 boates and a number of whalebotes, floating battery, Gondaloes, Rogalleyes and Gunbotes."

"The next day we arrived at the north end of Lake George and landed without opposition. The french that were encampt at that end of the Lake fled at our appearance as far as Ticonderoge and joined thare old commander Moncalm and we ware drawn up in order and divided into collams and ordered to March toward Moncalm in his camp before the fort—but unfortunately for us Moncalm like a Gineral dispatched five hundred to oppose us in our landing or at least to imbarres us in our March so he might put his camp in some sort of defense before our arme could arrive and thay did it most completely. We had not Marcht more than a mile and a half befoare we meat the falon (forlorn) hope for such it proved to be. The British troops kept (the) rode in one collam the Amaracans marcht threw ye woods on thare left. On ye rite of the British was the run of water that emteys from Lake George into Lake Champlain. The British and French meat in the open rode verey near each other befoar they discovered the french in a count of the uneaveneas of the ground. Lord How held the secont place in command and beaing at the head of the British troops with a small sidearm in his hand he ordered the troops to forme thare front to ye left to atack the french. But while this was dueing the french fird and his Lordship receaved a ball and three buck shot threw the senter of his brest and expired without spekeing a word. But the french pade dear for this bold atempt. It was but a short time befoare thay ware surounded by the hole of the Amaracan troops and those that did not leape into the rapid stream in order to regan thare camp ware made prisners or kild and those that did went down with the Raped curant and was drounded. From the best information I could geat from ye french of that partea was that thare was but seven men of ye five hundred that reacht the campt but it answered the purpas amaseingly."

"This afair hapend on thirsday. The troops beaing all strangers to the ground and runing threw the woods after the disparst frenchmen night came on and thay got themselves so disparst that thay could not find the way back to thare boates at the landing. That nite the British did beatter haveing the open rod to direct them thay got to ye lake Sid without trubel. A large party of ye amaracans past the nite within a Bout half a mile of the french lines without noeing whare thay ware til morning. I was not in this partey. I had wandered in ye woods in the nite with a bout twelve men of my aquantans—finealey fel on the

Rode a bout a mile north of ye spot whare the first fire began. Beaing in the rode we marched toward our boates at ye water side but beaing dark we made but a stumbling pece of bisness of it and sun coming among the dead bodeyes, which ware strewed quit thick on the ground for sum little distans. We stumbled over them for a while as long as thay lasted. At lengh we got to the water just before day lite in the morn."

"What could be found of the troops got in sum order and began our march a bout two a clock in ye afternoon crossing the raped stream and left it on our left the rode on this side was good and we advansd toward the french camp as far as the miles (mills) about a mile from the works and thare past the night lying on our armes."

"This delay gave the french what thay wanted—time to secure thare camp which was well executed. The next day which was Satterday about eleven we ware seat in mosin the British leading the van it was about. They ware drawn up before strong brest work but more in extent then to permit four thousand five hundred acting. We had no cannon up to the works. The intent was to march over this work but thay found themselves sadly mistaken. The french had cut down a grate number of pinetrease in front of thare camp at some distance. While som ware entrenching others ware imployed cuting of the lims of the trease and sharpening them at both ends for a shevo dufrease (chevaux de frise) others cuting of larg logs and geting them to the Brest works. At length thay ware ready for our resaption.

"About twelve the parties began thare fire and the British put thare plan on fut to march over the works but the lims and tops of the trease on the side for the diek stuck fast in the ground and all pointed at upper end that thay could not git threw them til thay ware at last obliged to quit that plan for three forths ware kild in the atempt but the grater part of the armey lade in the rear on thare fases til nite while the British ware batteling a brest work nine logs thick in som plases which was dun without ye help of canan tho we had as fine an artilrey just at hand as could be in an armey of fifteen thousand men but thay ware of no youse while thay ware lying on thare fases. Just as the sun was seating Abercrombie came from left to rite in the rear of the troops ingaged and ordered a retreat beat and we left the ground with about two thousand two hundred loss (actually over nineteen hundred) as I was informd by an officer who saw the returns of ye nite wounded and mising."

PETER POND: FUR TRADER & ADVENTURER

"We ware ordered to regain our boates at the lake side which was dun after traveling all nite so sloley that we fell asleep by the way. About nine or tenn in the morning we were ordered to imbark and cross the lake to the head of Lake George but to sea the confusion thare was the solgers could not find thare one botes but imbarked permisherley (promiscuously) whare ever thay could git in expecting the french at thare heales eavery minnet. We arrived at the head of the lake in a short time—took up our old incampment which was well fortefied."

"After a few days the armey began to com to themselves and found thay ware safe for the hole of the french in that part of the country was not more than three thousand men and we about fortee thousand. We then began to git up provishan from fort Edward to the camp but the french ware so bold as to beseat our scouting party between the camp and fort Edward and cut of all the teames, destroy the provishun, kill the parties and all under thare ascort. We past the sumer in that maner and in the fall verey late the camp broke up and what remaned went into winter qaters in different parts of the collanees thus ended the most ridicklas campane eaver hard of."

Pond's account of the campaign corroborates the general views which have been held by various writers. The clumsiness of General Abercromby, the loss of Lord Howe, and the fatal attack on the French without the support of artillery are the points stressed in explanation of the British defeat.

Pond had been engaged in two campaigns as a soldier on the Lake George—Lake Champlain front. In the first campaign he had endured the ennui of camp life and in the second campaign he had seen his first severe fighting with heavy casualties. He had no liking for a third campaign in this territory. "The year 59 an armey was rased to go against Niagaray to be commanded by Gineral (Prideaux). As the Connecticut troops ware not to be imploid in that part of the armey I went to Long Island and ingaged in thot sarvis"—joining the Suffolk County regiment on April 17, 1759. "In the Spring we repaired to Albany and gined the armey as that was the plase of Rondevuse. We ware imploid in geating forward provisons to Oswego for the sarvis of the Campain. When we asemled at Osawaga Col. Haldaman took part of the troops under his command and incampt on the Ontarey side but the troops that ware destind to go against Niagara incampt on the opaset side of the river under the command of General (Prideaux). But the

Company I belonged to was not ordered over the lake at all but Col. Johnston (Col. John Johnstone) who was in the Garsea (Jersey) Sarvis sent for me in partickler to go over the lake. I wated on him and inquired of him how he came to take me the ondley man of the Company out to go over the lake. He sade he had a mind I should be with him. I then asked him for as maney of the companey as would make me aseat of tent mates. He sun complid and we went and incampt with the troop for that sarvis. Capt. Vanvater (Thwaites suggests Capt. Van Veghte) commanded the company we joined."

"We sun imbarkt and arived at Nagarey. In a few days when all ware landead I was sent by the Agatint Mr. Bull as orderly sarjant to General (Prideaux). I was kept so close to my dutey that I got neither sleape nor rest for the armey was down at Johnsons landing four miles from the acting part of the armey. I was forced to run back and forth four miles nite and day til I could not sarve eney longer. I sent to Mr. Bull to releave me by sending another sargint in my plase which was dun and I gind my friends agane and fought in the trenches aganst the fort."

"Befoar we had capterd the fort the Gennarel had gind the arme and himself and my frend Col. Johnson ware both kilt in one day and Col. (Thodey) of the New York troops shot threw the leag. This was a loss to our small armey—three brave offesars in one day. We continued the seage with spereat under the command of Sir William Johnson who it fell to after the death of (Prideaux). I was faverd—I got but one slite wound dureing the seage. At the end of twenty-five days the fort capatalated to leave the works with the honners of war and lay down thare armes on the beach whare thay ware to imbark in boates for Schanactady under an escort. After apointing troops to garsen the fort we returned to Oswego and bilt a fort cald fort Erey." (Fort Ontario).

He began to realize that war meant the loss of friends and he writes,—"At the close of the Campain what was alive returned home to thare native plases but we had left a number behind who was in thare life brave men. On my arival at Milford I found maney of the prisners I had bin so industres in captering ware billeated in the town. I past the winter among them". Pond's knowledge of French was probably acquired at this time.

In the siege at Niagara and Pond's third campaign he had begun to impress his superior officers with his ability and was singled out for promotion. He had reached maturity. We find him writing "in 1760 I receaved a Commission and entered a forth time in the armey. We then gind the armey at the old plase of Rondavuse and after lying thare a few weakes in camp duing Rigimental dutey General Armarst (Amherst) sent of in pourshen to carre the baggage to Oswego whare part of the armey had all ready arived. I was ordered on this command—four offesers and eighty men. On our arrival at Oswego the Genarel gave the other three offesers as maney men as would man one boate and ordered them to return to thare rigiment. Me he ordered to incamp with my men in the rear of his fammerley til farther orders with seventy men til just befoar the armey imbarkt for S(wegatchie) and then gind my regiment. Sun after thare was apointed a light infantry companey to be pickt out of each regiment—hats cut small that thay mite be youneform. I was apointed to this Company."

"When orders ware given the armey about nine thousand imbark in a number of boates and went on the lake towards Swagochea whare we arived safe. Thare we found Pashoe (Pouchot) that had bin taken at Niagarey the sumer before commanding the fort and semed to be detarmined to dispute us and give us all the trubel he could but after eight or a few more days he was obliged to comply with the tarmes of our victoras armey a second time in les than one year."

"We then left a garrson and descended the river til we reacht Montreal the ondley plase the french had in possession in Canaday. Hear we lay one night on our armes. The next day the town suranderd to Gineral Amharst."

With the capture of Montreal the work of the army in the conquest of New France was finished. Peter Pond at the end of the campaigns was twenty years old. During the space of four years he had joined as a raw young recruit, had been engaged as a private in two campaigns, those of 1756 and 1758 on the Ticonderoga front, had become a sergeant at Niagara in 1759, and obtained his commission in 1760. He had become inured to the hardships of army life and to its discipline and had gained favour in the eyes of his superior officers. He acquired a knowledge of the army in a remarkably short space of time. He appears to have had an extraordinary physique. He had seen life at its worst but there stands out clearly in this part of his life his great loyalty to his friends. He rejoined the same company in 1758, he asked for a set of tent-mates to go with him to Niagara, he wanted to be returned to the ranks at Niagara, and he was a man who won his commission because of his ability to command men. These were impressionable years as one can gather from the detailed character of his description of the campaigns.

II. THE APPRENTICE TRADER

(1761-1773)

The conquest was over and for Peter Pond there came the problem of adjustment to civil life. "All Canaday subdued I thought thare was no bisnes left for me and turned my atenshan to the seas thinking to make it my profesion and in sixtey one I went a Voige to the islands in the West Indees and returned safe but found that my father had gon a trading voig to Detroit and my mother falling sick with a feaver dide (on June 16, 1761) before his return." For the time Pond found it necessary to curb his restless temperament which army experience had intensified. He writes "I was oblige to give up the idea of going to sea at that time and take charge of a young fammaley til my father returnd". After his father's return "I bent my mind after differant objects and tared in Milford three years". No information is given in his journal as to his activities but according to records he was married during this interval to Susanna Newell by whom he had at least two children one of whom, Peter Pond, the third, was born in 1763. These family ties explain his three year sojourn at Milford. "Which was the ondley three years of my life I was three years in one place sins I was sixteen years old up to sixtey".

Information on the activities of his life in the decade after 1763 is extremely scanty. His father apparently joined the numbers of traders in 1761 who rushed to the west after the conquest of New France. He may have returned in 1762 and doubtless found prospects had suffered through the uncertainties of Indian trade previous to the outbreak of Pontiac's war in 1763. It is known that he died insolvent in 1764 and that his principal creditors were Captain John Gibb and

Garrett Van Horn Dewitt. With the close of the Indian wars and possibly assuming his father's debts Peter Pond decided to follow in his father's footsteps. In 1765 he left Milford to engage in the Detroit trade. Little is known of his activities in the six years of his residence in the Detroit country. He writes in his journal "I continued in trade for six years in different parts of that countrey but beaing exposed to all sorts of Companey. It hapend that a parson (person) who was in trade himself to abuse me in a shamefull manner knowing that if I resented he could shake me in peaces at same time supposing that I dare not sea him at the pints or at leas I would not but the abuse was too grate. We met the next morning eairley and discharged pistels in which the pore fellowe was unfortenat. I then came down the country and declard the fact but thare was none to prosacute me". It is probable that like other traders he was under the close supervision which Sir William Johnson exercised over the southern posts. Along with other traders at Detroit, including Isaac Todd, he signed a petition[1] on 26th November, 1767, asking Sir William Johnson to restrict the amount of rum brought by traders to 50 gallons in each 3 handed battoe load of dry goods and also to permit the petitioners to trade beyond the fort of Detroit since otherwise the trade was lost to small unscrupulous traders. He appears to have wintered at some time at Michilimackinac since he mentions the fishing at the straits of Mackinac in winter. "I have wade a trout taken in by Mr. Camps with a hook and line under the ice in March sixtey six pounds wate." Again he writes "I was at Mackinac when Capt. George Turnbull comanded". Turnbull apparently went to Mackinac in 1770 and left for the West Indies with his regiment in 1773. Probably Pond was at Michilimackinac in the winter of 1770-1. His main interest during the period centred about Detroit, during the later years, extending his trade to outlying points as at Michilimackinac. With the extension of trade beyond the posts regulation became less effective and it was inevitable that Pond should meet numbers of rough characters who were notorious in the trade at that time. That Pond was not prosecuted for the results of the duel is evidence that these events were not uncommon.

---

[1] Sir William Johnson Papers, (Albany, 1927) Vol. V. p. 830. His headquarters probably continued at Detroit as a sale of 120 acres of land on Gros Point by Beaubien, a trader, for £200. 18s. 2d. N. Y. currency to "Peter Pond of Detroit, Merchant" is recorded on Aug. 13, 1770. Registrar des Notaries, Detroit, Vol. V. p. 105.

PETER POND: FUR TRADER & ADVENTURER

In 1771 he came down to Milford and made another "ture to ye West Indies" in 1772. On his return he received a letter from a Mr Graham in New York asking him to come down to arrange for a partnership to venture in the trade from Michilimackinac. This invitation was evidence that Pond's apprenticeship in the Detroit trade had ended. He had acquired wide experience in the trade and probably some capital. Graham had apparently been engaged in the Michilimackinac trade for several years as his name appears as an Albany trader at that point in 1767. In the intervening period he had probably acquired a substantial supply of capital from the trade by Green Bay to the Mississippi and in 1773 he appears to have been one of the largest traders to that district. It was significant of the impression which Pond had made on his fellow traders as to his ability as a trader, and his honesty and integrity, that he should have been chosen as the active partner for this important venture. No higher tribute could have been paid to him.

He had acquired at the end of this period a thorough grasp of the demands of the trade. He knew the character of goods required for a successful trade. He probably knew the Indian languages and had a knowledge of French. From his voyages to the West Indies and his army experience he had probably acquired a knowledge of astronomy and navigation sufficient to enable him to determine latitude and longitude. He had prepared himself for the active years which were to follow. He was a master trader at thirty-one.

III. THE MISSISSIPPI

A. THE TRADER (1773-1774)

A partnership with Graham was formed and a cargo weighing 4600 pounds was made up. It is probable that Graham furnished the major share of the capital and that Pond was the active partner. In 1773 at the age of thirty three Peter Pond entered his majority in the fur trade. In April he left Milford not to return for at least twelve years.

After the Conquest, traders from New York and Albany pushed westward to Detroit and Michilimackinac. New York became a more important depot for the fur trade at the expense of Montreal which had dominated the trade in the French regime. The advantages of New York were shown in the character of the route by which goods were sent by the Great Lakes to the interior. Pond gives a valuable description of the Albany route which he had reason to know thoroughly from his experience in forwarding supplies to Oswego in 1759 and 1760 and to Detroit in later years. Mr. Graham took the goods to Michilimackinac by this route but Pond was not less qualified to describe it. The goods were shipped from New York to Albany and freighted in wagons fourteen miles from Albany to Schenectady. At this point they were loaded on batteaus and hauled up the Mohawk River to Fort Stanwix. The batteaus and goods were hauled over land for the distance of one mile to Wood Creek. From Wood Creek they went through Oneida Lake down the Oswego river to Lake Ontario. The boats were taken along the south shore of Lake Ontario to the landing place at Niagara and hauled over the nine mile portage above the rapids and the falls to Fort Schlosser. From that point they were taken up the Niagara to Fort Erie and along the south side of the lake to Detroit across Lake St. Clair and along the west side of Lake Huron to Michilimackinac. Goods could be taken in comparatively large boats and the transport costs from New York by the Great Lakes favoured the development of trade from that point.

But Montreal had the advantage of experience in the trade such as had been gained in the long period of the French regime. Pond writes for example, "I wanted some small artickles in the Indian way to compleat my asortment which was not to be had in New York." Consequently he took a boat through Lake George and Lake Champlain to Montreal, "whare I found all I wanted". Again he was familiar with the route to the foot of Lake George from the unsuccessful attack on Ticonderoga. It was fifteen years since Pond had been engaged in the memorable battles around Lake George and thirteen years since he had been at the capitulation of Montreal. He was able to purchase the necessary goods and he found the fur trade already well organized under English auspices following the collapse of French trade after the Conquest. Several canoes were outfitting in the spring of 1773 for Michilimackinac some of them owned by the firm, of his old acquaintance of Detroit, Isaac Todd, and James McGill. He arranged with these two men to have his goods taken in their canoes and with them he embarked in a canoe from Lachine to Michilimackinac by the Ottawa River.

The route from Montreal by the Ottawa to Michilimackinac was new to Pond. It has been described in numerous fur trading journals in great detail. Pond apparently found little to interest him on the long up stream journey on the Ottawa to Lake Nipissing, down the French river to Georgian Bay, along the north shore of Lake Huron until they were opposite Mackinac, and across the strait to the island on which the British garrison was located. The incident of the trip which interested him greatly as it did other travellers was the ceremony at the church of St. Ann on the Lake of Two Mountains. The voyageurs deposited a small sum and "by that meanes thay suppose thay are protected" by St. Ann, their patron saint. No one but a fur trader would have noticed that "while absent the church is not locked but the money box is well secured from theaves". Pond arrived at Michilimackinac "where I found my goods from New York had arived safe".

Michilimackinac was the outfitting depot for traders going to winter in the interior around Lake Michigan, Green Bay, and the Mississippi, or around Lake Superior. "Thare was a British Garason whare all the traders assembled yearley to arang thare afaires for the insewing winter". "Hear I met with a grate meney hundred people all denominations". At various periods Pond spent considerable time at this depot and his description of Mackinac is worthy of addition to the large number of descriptions already in existence. He begins "This place is kept up by a Capts. command of British which were lodged in good barracks within the stockades whare thare is some french bildings and a commodious Roman church whare the French inhabitants and Ingasheas (engagés) go to mass". Pond was much impressed with the French Catholics and their religion. "Befoare it was given up to the British thare was a French Missenare astablished hear who resided for a number of years hear. While I was hear thare was none but traveling one who Coms sometimes to make a short stay but all way in the Spring when the people ware ye most numeras then the engashea often went to confes and git absolution". The district had a large floating population, and Pond was impressed by the number of people who had come from Quebec. "Most of the frenchmen's wives are white women". "The inhabitans of this plase trade with the natives and thay go out with ye Indians in the fall and winter with them—men, women and children.—In the Spring thay make a grate quantity of maple sugar for the youse of thare families and for sale som of them". Michilimackinac was primarily a trading locality. "The land about Macinac is vary baran—a mear sand bank—but the gareson by manure have good potaters and sum vegetables. The British cut hay anuf for thare stock a few miles distans from the gareson and bring hom on boates. Others cut the gras and stock it on the streat (strait) and slead it on the ice thirty miles in ye winter". Fish was important as a staple food, and the traders were to a large extent dependent on this resource. Pond writes "I have wade a trout taken in by Mr. Camps with a hook and line under the ice in March sixtey six pounds wate. I was present. The water was fifteen fatham deape. The white fish are ye another exquisseat fish. They will way from 2-1/2 to 9 and 10 pound wt."

Alexander Henry has described the village of the Ottawas at L'Arbre Croche from which he obtained supplies of Indian corn and sugar. Pond notes the existence of these Indian villages some twenty or thirty miles distant "whare the natives improve verey good ground. Thay have corn beens and meney articles which thay youse in part themselves and bring the remainder to market. The nearest tribe is the Atawase and the most sivelised in these parts but drink to exses. Often in the winter thay go out on a hunting party. In ye Spring thay return to thare villages and imploy the sumer in rasein things for food as yousal.

PETER POND: FUR TRADER & ADVENTURER

But this is to be understood to belong to the women—the men never meadel—this part of thare bisness is confind to the females ondley. Men are imployd in hunting, fishing and fouling, war parties etc. These wood aford partreages, hairs, vensen fixis and rackcones, sum wild pigins". During the summer tribes from the surrounding territory and from "a grate distance" came to Michilimackinac to trade, bringing furs, skins, maple sugar, "dride vensen, bares greas and the like which is a considerable part of trade."

Pond's immediate task at Michilimackinac was that of rearranging the goods which had arrived by boats, purchasing canoes, provisions and supplies, and hiring men and loading the bales in the canoes. Equipments were made up for the different parts of the country. He divided his goods into twelve parts and "fited out twelv larg canoes for differant parts of the Mississippy river. Each cannew was mad of birch bark and white seader thay would carry seven thousand wate". "In Sept. I had my small fleat readey to cross lake Mishegan".

Pond had engaged nine clerks to handle the separate outfits to different parts and with this large contingent he left for Green Bay. He reached the mouth of the bay in three or four days and crossed over to the southwest side in the lea of some islands. On this side they followed the shore to the mouth of the river and to the small French farming village a short distance up. The people raised corn and "sum artikles for fammaley youse in thare gardens" and they had "fine black cattal and horses with sum swine". They traded to a certain extent with the Indians going by this route. Pond also mentions the Menominee village on the north side of the Bay, the people of which were chiefly hunters although they depended also on wild rice which they gathered in September.

After two days in the French village they ascended to the Puan village at the east end of a lake Winnebago[2]. The women raised corn, beans and pumpkins and they lived on rabbits, partridges and some venison but there was little fish in the lake. Pond was not impressed with these Indians and "we made but a small stay". He was not conversant with their language. "They speake a hard un couth langwidge scarst to be learnt by eney people". He narrates with great zest the story of the visit of a chief and a small band of this nation to Capt.

---

[2] L. P. Kellogg, *The French regime in Wisconsin and the Northwest* (Madison, 1925) p. 314 a map dated 1730.

Turnbull at Michilimackinac "He held a counsel with them but he couldn't get an intarpetar in the plase that understood them. At length the Capt. said that he had a mind to send for an old Highland solge that spoke little but the harsh langwege—perhaps he mite understand for it sounded much like it". They were not a sociable tribe—"Thay will not a sosheat with or convars with the other tribes nor inter-marey among them—They live in a close connection among themselves". Pond's curiosity was aroused by these people as he later enquired at Detroit "of the oldest and most entelagant Frenchman" regarding them. He was told that they formerly lived west of the Missouri, that they were very quarrelsome and that they were driven by other tribes across the Missouri and the Mississippi to their present location. The Fox tribe it was thought lived near them, fitting neighbours, since this tribe was driven from Detroit for misbehaviour. Whether this reputation was deserved or whether Pond had tried to carry on trade with them without success or whether he had traded with them and lost, it would be difficult to determine but the chief objection to them was given forcibly. "They are insolent to this day and inclineing cheaterey thay will if thay can get creadit from the trader in the fall of ye year to pay in the spring after thay have made thare hunt but when you mete them in Spring as know them personeley ask for your pay and thay will speake in thare one language if thay speake at all which is not to be understood or other ways thay will look sullky and make you no answer and you loes your debt", which sounds like the voice of sad experience.

    Later acquaintance did not improve early impressions. Pond continued up the river to the Grand Butte des Morts where this tribe "yous to entar thare dead when thay lived in that part". He describes the ceremony of some of the natives gathered to pay their respects to one of the departed. "Thay had a small cag of rum and sat around the grave. Thay fild thar callemeat (calumet) and began thar saremony by pinting the stem of the pipe upward—then giving it a turn in thare and then toward ye head of the grav—then east and west, north and south after which thay smoked it out and filf it agane and lade (it) by—then thay took sum rum out of the cag in a small bark vessel and pourd it on the head of the grave by way of giving it to thar departed brother—then thay all drank themselves—lit the pipe and seamed to enjoi themselves verey well. Thay repeated this till the spirit began to operate and thare harts began to soffen. Then thay began to sing a song or two but at the end of every song thay soffened the clay. After sumtime had relapst the cag had bin blead often. Thay began to repete the satisfaction thay had with that friend while he was with them and how fond he was of his frends while he could git a cag of rum and how thay youst to injoy it togather. They amused themselves in this manner til they all fell a crying and a woful nois they made for a while til thay thought wisely that thay could not bring them back and it would not due to greeve two much—that an application to the cag was the best way to dround sorrow and wash away greefe for the moshun was soon put in execution and all began to be marey as a party could bee. Thay continued til near nite. Rite wen thay ware more than half drunk the men began to aproach the females and chat frelay and apearantley friendly. At lengh thay began to lean on each other, kis and apeared verey amaras.—I could observe clearley this bisiness was first pusht on by the women who made thare visit to the dead a verey pleasing one in thare way. One of them was quit drunk as I was by self seating on the ground observing thare saremones, cam to me and askt me to take a share in her bountey—But I thought it was time to quit and went about half a mile up the river to my canoes whare my men was incampt but the Indians never came nigh us. The men then, shun (mentioned) that three of the women had bin at the camp in the night in quest of imploy." With these observations Pond left this tribe of Indians.

His trip over the portage from the St. Lawrence drainage basin to the Mississippi drainage basin by the Fox River into the Wisconsin river was the next object of concern. The whole river was easily navigated as far as the lake of the Puans and except for one or two small rapids from the lake as far as the portage. Above the rapids the channel became narrow and very winding and the water less swift. "In maney parts in going three miles you due not advans one. The bank is almost leavel with the water and the medoes on each sid are clear of wood to a grate distans and clothd with a good sort of grass the openings of this river are cald lakes but thay are no more than larg openings. In these plases the water is about four or five feet deap. With a soft bottom these plases produce the gratest quantatys of wild rise of which the natives geather grat quantities and eat what thay have ocation for and dispose of the remainder to people that pass and repass on thare trade. This grane looks in its groth and stock and ears like ry and the grane is of the same culler but longer and slimer. When it is cleaned fit for youse thay boile it as we due rise and eat it with bairs greas and sugar but the greas thay ad as it is bileing which helps to soffen it and make it brake in the same maner as rise. When thay take it out of thare cettels for yous thay ad a little sugar and is eaten with fresh vensen or fowls, we yoused it in the room of rise and it did very well as a sub-statute for that grane as it busts it turns out perfeckly white as rise". Wild rice and ducks were of appreciable value in conserving the food supply. The ducks had fattened on the wild rice and "when thay ris made a nois like thunder. We got as meney as we chose fat and good.—You can purchis them verey cheape at the rate of two pens per pese. If you parfer shuting them yourself you may kill what you plese".

After Pond's escape from the Indian women, he and his party proceeded up the winding river to a shallow lake with an abundance of wild rice and ducks. They encamped here for the night and spent "the most of ye next day to get about three miles—with our large cannoes the track was so narrow. Near nite we got to warm ground whare we incampt and regaled well after the fateages of the day". On the next day more slack water and a winding stream "we have to go two miles without geating fiftey yards ahead so winding" but at night they came within sight of the portage and arrived there at noon next day. They unloaded the canoes and "toock them out of the water to dry that thay mite be liter on the caring plase."

He describes the portage as very level for two thirds of a mile and consequently bad in wet weather. In the centre the ground rises to a considerable height and is covered with a fine open wood similar to that located back from the banks of the river. The land was described as excellent and covered with good timber,—the fires having destroyed the small wood. The height of land located in the centre extended for about one quarter of the whole distance. "The south end is low flat and subject to weat." "After two days hard labor we gits our canoes at the carring plase with all our goods and incampt on the bank of the river Wisconstan and gumd our canoes fit to descend that river. About midday we imbarkt."

Pond's journal represents with full force his reactions to the variety of travel. Writing from memory the physical reactions had left a strong impression. He notes with feeling "the warm ground" on which to camp after a hard day in marshy country, "the fat and good" wild ducks. After crossing the portage he describes the river in two places as "a gentel glideing stream". "As we desended it we saw maney rattel snakes swimming across it and kild them."

In about a day's travel they reached the village of the Sauks on the north side of the river. At this village he stayed two days "This beaing the last part of Sept. there people had eavery artickel of eating in thare way in abundans". They were of interest to Pond. "Thay are of a good sise and well disposed—les inclind to tricks and bad manners than thare nighbers. Thay will take of the traders goods on creadit in the fall for thare youse. In winter and except for axedant thay pay the deat (debt) verey well for Indians I mite have sade inlitend or sivelised indans which are in general made worse by the operation".

To the anthropologist the Wisconsin portage route had the greatest interest and the fur trader was always an anthropologist. On this route a number of tribes had settled partly because of its importance to trade and the incidental wars which had swept the area in the French period and partly because of its position—marginal to the plains Indians, the southern woods Indians, and the northern Indians. Pond noted the cultural traits of the Sauks. In the first place "thare villeag is bilt cheafely with plank thay hugh out of wood—that is ye uprite—the top is larch (arched) over with strong sapplins sufficient to suport the roof and covered with barks which makes them a tile roof. Sum of thare huts are sixtey feet long and contanes several fammalayes. Thay rase a platform on each side of thare huts about two feet high and about five feet broad on which thay seat and sleap. Thay have no flores but bild thare fire on the ground in the midel of the hut and have a hole threw the ruf for the smoke to pas".

The total population was about one hundred. "The women rase grate crops of corn, been, punkens, potatoes, millans and artikels". They had a comparatively large cleared space with excellent land. Fishing was very poor, "wild foul thay have but few", "Thare woods afford partrageis, a few rabeat, bairs and deear are plenty in thare seasons". In the fall "thay leave thare huts and go into the woods in quest of game and return in the spring—before planting time." "The men often join war parties with other nations and go against the Indans on the Miseure and west of that. Sometimes thay go near St. Fee in New Mexico and bring with them Spanish Horseis."

"Thare amusements are singing, dancing, smokeing, matcheis, gameing, feasting, drinking, playing the slite of hand, hunting, and thay are famas in Mageack. Thar religion is like most of the tribes. Thay alow thare is two sperits—one good who dweles a bove the clouds superintends over all and helps to all the good things we have and can bring sickness on us if he pleases—and another bad one who dweles in the fire and air, eavery whare among men and sumtimes dose mischief to mankind"—in other words they had come under the influence of French missionaries.

Like most traders Pond was interested in their courtship and marriages. "At night when these people are seating round thare fires the elderly one will be teling what thay have sean and heard or perhaps thay may be on sum interesting subject. The family are lisning. If thare be aney young garl in this lodg or hut that aney man of a differant hut has a likeing for he will seat among them. The parson of his arrant (errand) being prasent hea will watch an opertunity and through a small stick at hair (her). If she looks up with a smile it is a good omen. He repets a second time perhaps ye garle will return the stick. The Semtam (sympton) ar still groing stronger and when thay think proper to ly down to slape each parson raps himself up in his one blanket. He taks notes whar the garl seats for thare she slepes. When all the famaley are quiet a(nd) perhaps a sleap he slips soffely into that and seat himself down by her side. Presantlay he will begin to lift her blanket in a soft maner. Perhaps she may twich it out of his hand with a sort of a sie and snore to gather but this is no kiling matter. He seats awhile and makes a second atempt. She may perhaps hold the blankead down slitely. At lengh she turns over with a sith and quits the hold of the blanket.—This meatherd is practest a short time and ye young Indan will go ahunting and (if) he is luckey to git meat he cum and informs the famaley of it and where it is he brings the lung and hart with him and thay seat of after the meat and bring it home this plesis and he begins to gro bold in the famerley. The garl after that will not refuse him.—He will then perhaps stay about the famarley a year and hunt for the old father but in this instans he gives his conseant that thay may sleap togather and when thay begin to have children thay save what thay can git for thare one youse and perhaps live in a hut apart".

As in most visits described by traders the voyageurs lost little time in making the acquaintance of the women. Pond remarks of this tribe "Thay are not verey gellas of thare women".

"After I had given them a number of cradeat to receve payment the next spring I desended to the fox villeag—about fiftey miles distans." Again like most traders Pond attributes the maliciousness of this tribe (Foxes) to the influence of the missionaries. "Hear I meat a differant sort of people who was bread at Detroit under the french government and clarge (clergy) till thay by chrisanising grew so bad thay ware oblige to go to war against them". He heard echoes of the

Fox wars[3] of the French regime in which the Fox Indians were driven from Detroit and forced to flee to the Fox river, and in which the French again carried on war against them to break up their monopoly over the Fox river portage, and he attributed their misfortunes in part to these wars. But the Foxes were also suffering from disease. "As I aprocht the banks of the villeag I perseaved a number of long Painted poles on which hung a number of artickels, sum panted dogs and also a grate number of wampum belts with a number of silver braslets and other artickels in the Indan way. I inquired the cause. Thay told me thay had a shorte time before had a sweapeing sickness among them which had caread of grate numbers of inhabitans and thay had offered up these sacrafisces to apease that being who was angrey with them and sent the sickness—that it was much abated tho thar was sum sick. Still I told them thay had dun right and to take cair that thay did not ofend him agane for fear a grater eavel myte befall them". The produce of their fields was reduced as a result of the sickness[4] and Pond stayed only one day. He got the articles he needed and which they could spare and "gave them sum creadeat and desended the river to the mouth which emteys into the Masseippey and cros that river and incampt".

There follows no rhapsody on the mighty Mississippi. "Just at night as we ware incampt we perseaved large fish cuming on the sarfes of the water". After leaving the Fox village Pond was apparently joined by another trader with several men—in any case he camped with another trader. Hooks and lines were put out in the river and in the morning they hauled in the catch. "They came heavey. At lengh we hald one ashore that wade a hundered and four pounds—a seacond that was one hundred wate—a third of seventy five pounds.—The fish was what was cald the cat fish. It had a large flat head sixteen inches between the eise.—The men was glad to sea this for thay had not eat mete for sum days nor fish for a long time. We asked our men how meney men the largest would give a meale. Sum of the largest eaters sade twelve men would eat it at a meal. We agreed to give ye fish if thay would find twelve men that would undertake it. Thay began to dres it.—They skind it—cut it up in three larg coppers such as we have for the youse of our men. After it was well boild thay sawd it up and

---

[3] See L. P. Kellogg, *op. cit.*, chs. XIII, XV.
[4] Carver notes an epidemical disorder among this tribe in 1766. *Travels* (London, 1781) p. 48. His journal should be compared with that of Pond 8 years later.

all got round it. Thay began and eat the hole without the least thing with it but salt and sum of them drank of the licker it was boild in. The other two was sarved out to the remainder of the people who finished them in a short time. Thay all declard thay felt the beater of thare meale nor did I perseave that eney of them ware sick or complaind".

The following morning they crossed the river and ascended about three miles "to the Planes of the dogs" (Prairie du chien). The plain was a large level stretch of land on the east side of the river at the junction of the Wisconsin and the Mississippi. Prairie du Chien was the rendezvous of traders and Indians from New Orleans, from Michilimackinac, and the tributaries of the Mississippi. When Pond arrived he "meat a larg number of french and Indans making out thare arrangements for the insewing winter and sending of thare cannoes to differant parts—likewise giving creadets to the Indans who ware all to Rondoveuse thare in spring. I stayed ten days sending of my men to different parts." His nine clerks were dispatched to various tributaries.

After completing the business at Prairie du Chien, Pond and two other traders left in October for St. Peters River. They proceeded very slowly lest they should overtake the Nottawaseas who had preceded them and in order that "we mite not be trubeld with them for creadit as thay are bad pay masters"—"We had plenty of fat gease and ducks with venson—bares meat in abundans—so that we lived as well as hart could wish on such food—plentey of flower, tea, coffee, sugar and buter, sperits and wine, that we faird well as voigers. The banks of ye river aforded us plenty of crab apels which was verey good when the frost had tuchd them at a sutabel tim". Eventually they reached St. Peters river and Pond claims they found about fourteen miles from the mouth Carver's old hut in which he had wintered in 1766-7. "It was a log house about sixteen feet long covered with bark—with a fireplase but one room and no flore".

Finally they decided to build their houses for the winter. "We incampt on a high bank of the river that we mite not be overflone in the spring at the brakeing up of the ice, and bilt us comfortbel houseis for the winter and trade during the winter and got our goods under cover.—In Desember the Indans sent sum young men from the planes a long the river to look for traders and thay found us. After staying a few days to rest them thay departed with the information to thare frends. In Jany. thay began to aproach us and brot with them drid and grean meat, bever, otter, dear, fox, woolf, raccone and other skins to trade. They ware welcom and we did our bisness to advantage." Pond had a French competitor "for my nighber who had wintered among the Nottawase several winters in this river well knone by the differant bands. I perseaved that he seamed to have a prefrans and got more trade than myself. We ware good frends. I told him he got more than his share of trade but obsarved at ye same time it was not to be wondered at as he had bin long a quanted. He sade I had not hit on ye rite eidea. He sade that the Indans of that quorter was given to stealing and aspachely the women. In order to draw custom he left a few brass things for the finger on the counter—sum needles and awls which cost but a trifel, leattel small knives—bell and such trifels. For the sake of stealing these trifels thay com to sea him and what thay had for trade he got. I beleaved what he sade and tried the expereament—found it to prove well after which I kept up sides.—We proseaded eastward with ease and profet till spring".

The ice finally broke up in the spring and according to Pond the water rose 26 feet washing away a large part of the banks. With the fall of the water they loaded their canoes and drifted down stream to Prairie du Chien—"whare we saw a large colection from eavery part of the Misseppey who had arived before us—even from Orleans eight hundred leagues belowe us."—The boats from New Orleans "are navagated by thirty-six men who row as maney oarse. Thay bring in a boate sixety hogseats of wine on one—besides ham, cheese etc.—all to trad with the french and Indans." Fall and spring Prairie du Chien was the scene of great activities in the fur trade. "The Indians camp exeaded a mile and a half in length. Hear was sport of all sorts". "The gratest games are plaid both by french and Indians. The french practis billiards—ye latter ball." But Pond was a trader "we went to collecting furs and skins—by the differant tribes with sucsess". Competition was apparently keen with numerous French traders. He estimated not less than 130 canoes from Michilimackinac each carrying sixty to eighty hundredweight as well as numerous boats from New Orleans, Illinois and other parts, and that 1500 hundred pound packs went to Mackinac. "All my outfits had dun well. I had grate share for my part as I furnish much the largest cargo on the river. After all the bisness was dun and people began to groe tirde of sport, thay began to draw of for thare differant departments and prepare for the insewing winter. In July I arived at Mackenaw whare I found my partner Mr. Graham from New York with a large cargo. I had dun so well that I proposed to bye him out of ye cosarn and take it on myself. He excepted and I paid of the first cargo and well on towards the one he had brot me". Here ended the first venture.

From a pecuniary point of view Pond had learned the lessons of the fur trade to advantage. He knew men and was able to judge men to superintend his outfits. Success in the trade was dependent on judgment of character since the men were far removed from direct supervision throughout the winter.

He knew how to get along with competing traders and his old genius for making friends stood him in good stead. For a competitor to give information enabling him to get a larger share of trade is no slight testimony to the man's character. To be able to gather information as to the reputations of various Indians in repaying their debts required a genial disposition as well as shrewd judgment. He had camped, travelled and traded with competitors and with pronounced success. He also knew how to handle his voyageurs. "I had with me one who was adicted to theaving—he took from me in silver trinkets to the amount of ten pound but I got them agane to a trifle." And he emerged with sufficient profit to warrant a venture on his own account.

B. THE PEACEMAKER (1774-1775)

Pond had arrived at Michilimackinac, had disposed of his cargo and was immediately engaged in acquiring a new outfit. "I apleyd myself closely to ward fiting out a cargo for the same part of the country". With the coming of the traders Michilimackinac was similar to Prairie du Chien. "Hear was a grate concors of people from all quorters sum preparing to take thair furs to Cannadey—others to Albaney and New York—other for thare intended wintering grounds—others trade in with the Indans that come from different parts with thare furs, skins, sugar, grease, taller etc.—while others ware amuseing themselves in good company at billiards, drinking fresh french wine and eney thing thay please to call for while the more vulgar ware fiteing each other. Feasting was much atended to—dancing at nite with respectabel parsons. Notwithstanding the feateages of the industress the time past of agreabley for two months when the grater part ware ready to leave the plase for thare differant wintering ground".

The cargo and canoes had to be purchased and the men to be hired. It would be difficult to state the number of Pond's men who rehired with him but he probably had little difficulty in securing a complement. The young man Baptiste who had committed the theft during the winter almost certainly rehired. At Michilimackinac one of the visiting priests was hearing confessions and "the young man heard from his comrads who had bin to confess" of the priest "who was doing wonders among the people". "His consans smit him and he seat of to confess but could not get absolution. He went a seacond time without sucksess but was informed by his bennadict that something was wanting. He came to me desireing me to leat him have two otter skins promising that he would be beatter in future and sarve well. I leat him have them. He went of. In a few minets after or a short time he returned. I askt him what sucksess. O sade he the father sais my case is a bad one but if I bring two otter more he will take my case on himself and discharge me. I let him have them and in a short time he returned as full of thanks as he could expres and sarved me well after". Pond had acquired a full complement and he writes "I had now a large and rich cargo".

At this point trouble appeared. About the first of August a trader coming from Lake Superior brought news that war had broken out between the Sioux and the Chippewas, "and made it dangres for the trader to go in to the country". The only prospect of peace lay in the hands of the commander of the garrison. "A counsel was cald of all the traders and the commander laid his information befoar the counsel and told them it was out of his power to bring the government into eney expens in sending to these but desird that we would fall on wase and means among ourselves and he would indeaver to youse his influens as commanding offeser. We heard and thanked him we then proseaded to contrebute towards making six large belts of wampum— three for the Notawaysease and thre for the Ochpwase. Thay ware compleated under the Gidans of the comander and speacheis rote to both nations. I was bound to the senter of the Notawaseas contrey up St. Peters river. The counsel with ye commander thought proper to give me ye charge of thre belt with the speacheis and the traders to Lake Superer ware charged with the others. The import of the bisness was that I should send out carrears into the planes and—all the chefes to repare to my tradeing house on the banks of St. Peters river in the spring and thare to hear and obsarve the contents of the offesers speache and look at the belts and understand thare meaning—likewise to imbark and acompany me to Mackenac". Pond assumed the role of peace-maker as well as trader and the lives of many as well as of the traders depended on his success.

We are told nothing of his second journey from Michilimackinac to Prairie du Chien, but it probably did not differ from his first. At Prairie du Chien he found the Indians disturbed but not about the war. On inquiring "thay gave me to understand thare was a parson at that plase that had an eevel sperit. He did things beond thare conseption. I wished to sea him and being informed who he was I askd him meney questions. I found him to be a french man who had bin long among the nations on the Misura that came that spring from the Ilenoas" to Prairie du Chien. "He had the slite of hand cumpleately and had such a swa(y) over the tribes with whom he was aquanted that thay consented to moste of his requests. Thay gave him the name of Minneto (Manitou) which is a sperit.—As he was standing among sum people thare came an Indan up to them with a stone pipe or callemeat carelessly rought and which he seat grate store by—It was three times larger than minnetos mouth"—and "made of the read stone of St. Peters river so much asteamed among the eastern and southern nations".—"Minneto askd ye Indian to leat him look at it and he did so. He wished to purchis it from the Indan but he would not part with it. Minneto then put it into his mouth as the Indan supposed and swallod it. The poor Indan stood astonished. Minneto told him not to trubel himself about it—he should have his pipe agane in two or three days—it must first pass threw him. At the time seat the pipe was presented to the Indan. He looked upon it as if he could not bair to part with it but would not put his hand upon it. Minneto kept the pipe for nothing".

After enjoying this episode he left Prairie du Chien and reached his old house on St. Peters river on the thirteenth day. Pond was now actively interested in prosecuting the trade and he knew the country and the people. He learned from other Indians of a large band about 200 miles farther up "which wanted to sea a trader. I konkluded ameatley to put a small asortment of goods into a cannoe and go up to them—a thing that never was atempted before by the oldest of the traders on acount of the rudeness of those people who ware Nottawaseas by nation but the band was called Yantonoes—the cheafe of the band allwase lead them on the plaines". Pond writes that the Sioux were formerly one nation but they had broken up because of internal disputes into "six different bands each band lead by chefes of thare one chois". These bands were "1. the Yantonose (Yankton branch of the Dakotas or Sioux) 2. the band of the leaves. (the Wapeton branch) 3. the band of the wes. (the Leaf buds one of the divisions of the Santee) 4. the band of the stone house. (the Sisseton division of the Dakota) the other two bands are north one cald Assonebones (assiniboines) the others dogs ribs. These—speke the same langwege at this day".

    The anxiety with which the Yanktons awaited the arrival of a trader was shown by the arrival of the chief to invite him to go up and trade just as he was preparing to leave. They both set off together, the chief going by land and Pond by water. The chief arrived before him since he was able to take a more direct route "across the plaines". "I was nine days giting up to thare camp". By this time it was getting late in the season. Coming to a "larg sand flat by the river side"—"about three miles from the Indians camp and it beaing weat wether and cold I incampt and turned up my canoe which made us a grand shelter. At night it began to snow and frease and blowe hard.—Earley in the morning the wind took the canew up in the air—leat hir fall on the frozen flat and broke hir in pecis". Fortunately a number of the Indians came down on the opposite side of the river about noon, crossed over and "offerd me thare asistans to take my goods to thare camp. I was glad and excepted thare offer".

    The horses were loaded and the party marched to the camp. This was Pond's first experience in meeting Indians who were strangers to white traders. "Thay never saw a trader before on thare one ground at least saw a bale of goods opend. Sum traders long before sent thare goods into the planes with thare men to trade with these people—thay often would have them cheaper than they french men could sell them. These people would fall on them and take ye goods from them at thair one price til thay could not git eney. I was the first that atempted to go thare with a bale of goods".

    His reception by them involved numerous ceremonies. "Cuming near the camp made a stop and seat down on the ground. I preseaved five parsons from the camp aproching—four was imployd in caring a beaver blanket finely panted—the other held in his hand a callemeat or pipe of pece—verey finely drest with different feathers with panted haire. They all seat by me except the one who held the pipe. Thay ordered the pipe lit with a grate dele of sarremoney. After smoeking a fue whifs the stem was pinted east and west—then north and south—then upward toward the skies—then to ye earth after which we all smoked in turn and apeard verey frendlye. I could not understand one word thay said but from thare actions I supposed it to be all frendship. After smoking thay toock of my shoes and put on me a pair of fine Mockasans or leather shoes of thare one make raught in a cureas manner—then thay lade me down on the blanket—one hold of each corner and cared me to the camp in a lodg among a verey vennarable asembly of old men. I was plased at the bottom or back part which is asteamed the highest plase. After smoking an old man ros up on his feet with as much greaveaty as can be conseaved of he came to me—laid his hands on my head and grond out—I—I—I three times—then drawed his rite hand down on my armes faneing a sort of a crey as if he shead tears—then sit down—the hole follode the same exampel which was twelve in number. Thare was in the midel of the lodg a rased pece of ground about five inches in hight five feet long two and a half brod on which was a fire and over that hung three brass kettles fild with meete boiling for a feast.—At length an old man toock up some of the vittels out of one of ye kittles which apeared to be a sort of soope thick and with pounded corn mele. He fead me with three sponfuls first and then gave me the dish which was bark and the spoon made out of a buffeloes horn to fead myself. As I had got a good apatite from the fateages of the day I eat hartey.— While we ware imployd in this saremony thare was wateing at the dore four men to take me up and care me to another feast".—As soon as he had finished the first feast "The people in wateing then toock me and laid me on another skin and carred me to another lodg whare I went threw the same sarremony. There was not a woman among them— then to a third after which I was taken to a large (lodge) prepaird for me in which they had put my people and goods with a large pile of wood and six of thare men with spears to gard it from the croud".

Pond had been advised by the chief who had come down the river to invite him to come up to trade with the Indians that the trade "was to begin at sundown", and he had warned him that "if I was to contend with them thay mite take all that I had". Unfortunately the chief was absent and "thay compeld me to begin befoar the time—At four o'clock I cummenced a trade with them but ye croud was so grate that the chefe was obliged to dubel this gard (of six) and I went on with my trade in safety—seventy five loges at least ten parsons in each will make seven hundred and fifty. My people ware by-standers—not a word to say or acte".—"I was in a bad sittuation but at sundown the chefe arived and seeing the crowd grate he put to the gard six men more and took the charge on himself. He was as well obade and kept up as smart disapline as I ever saw. One of ye band was more than commonly dairing—he ordered one of the gard to throw his lans threw him in case he persisted in his impedens—the fellow came again—the sentanal threw his lans and it went threw his close and drew a leattel blod but he neaver atempted agane. I continued my trade till near morning. By that time thare furs ware gon".

"Thay prepared to March of as thay had lane on the spot sum time befour my arival thay had got out of Provishon. I was not in a situation to asist them beaing destatute myself". But Pond had still to fulfil his mission as peace-maker. "I informd the chefe of the belts I had with me and ye commanding offisers speach and desird him to make a speach befour thay decampt. This chefes name was Arechea. The chefe that came to me first had a smattran of the Ochipway tung—so much so that we understood each other at least suffisantly to convarse or convae our idease. He made a long speach. By the yousel sine of a shout threw the camp thay ware willing to cumply". They began to move off. "By day lite I could not sea one but the chefe who cept close by me to the last to prevent aney insult which mite arise as thay ware going of.—These people are in thare sentaments verey averishas but in this instans thay made not the least demand for all thare sarvis. Late in the morning the chefe left me".

In spite of the shortness of the time Pond had seen this band of Yankton Indians he had learned much about their life and customs. Undoubtedly his knowledge was supplemented by information gained from other tribes but he was a keen observer. He notes that they "are faroshas and rude in thare maners perhaps being in sum masher to thare leadg an obsger (obscure) life in the planes. Thay are not convarsant with evrey other tribe. Thay seldom sea thare nighbers. They leade a wandering life in that extensive plane betwene the Miseura and Missicippey". "The planes where these people wander is about four hundred miles brod east and west three hundred north and south". "Thay dwell in leather tents cut sumthing in form of a Spanish cloke and spread out by thirteen in the shape of a bell—the poles meet at the top but the base is forten in dimerter—thay go into it by a hole cut in the side and a skin hung befour it by way of a dore—thay bild thare fire in the middel and do all thare cookery over it—at night thay lie down all around the lodg with thare feat to the fire—Thay make youse of Buffeloes dung for fuel as there is but little or no wood upon the planes."—"Thay have a grate number of horses and dogs which carres there bageag when thay move from plase to plase."—"When thay are marching or riding over the planes thay put on a garment like an outside vest with sleves that cum down to thare elboes made of soft skins and several thicknesses that will turn an arrow at a distans—and a target two and a half feet in diameter of the same matearel and thickness hung over thare sholders that gards thare backs. When there is a number of them to gather going in front of thare band thay make a war like apearans.—Thay are continuely on the watch for feare of beaing sarprised by thare enemise who are all round them. Thare war implements are sum fire armes, boses and arroes and spear which thay have continuely in thare hands. When on the march at nite thay keep out parteas on the lookout".

These Indians had cultural traits of peoples living on the plains and dependent on the buffalo. "Thay run down the buffelow with thare horses and kill as much meat as thay please. In order to have thare horseis long winded thay slit thair noses up to the grissel of thare head which makes them breath verey freely. I have sean them run with those of natrall nostrals and cum in apearantley not the least out of breath".

From the standpoint of the higher priced furs the country was not promising. "Thay make all thare close of differant skins. These parts produce a number of otters which keep in ponds and riveleats on these planes and sum beavers but the land anamels are the mane object (of) the natives".

As to other traits Pond is not less observant. "Thay are verey gellas of thare women.—Thay genaley get thare wife by contract with the parans.—It sumtimes happens that a man will take his nighbers wife from him but both are oblige to quit the tribe thay belong to but it is seldum you can hear of murders cummitted among them. They have punnishment for thefts among themselves. They sumtimes retelate by taking as much property from the ofender if thay can find it but I seldum hurd of thefts among themselves whatever thay mite due to others.—Thay beleve in two sperits—one good and one bad.—When a parson dies among them in winter thay carrea the boddey with them til thay cum to sum spot of wood and thay put it up on a scaffel till when the frost is out of the ground thay intare it".

Meanwhile the Yanktons had departed. "I went to work bundling or packing my furs which I got from them. I was now destatute of frends or assistans except my one men and thay could not aford me aney asistans in the provishon line of which I was much in want. Nighther could thay assist me in the transportation of my furs. I then concluded to leave a boy to take care of them until we could return with sum provishon. The poor fellow seamd willing to stay by himself and all we could aford him was three handfulls of corn. In case of want I left him two bever skins which had sum meat on them and wone bever skin which he could singe the haire of and roste in the fire that he mite live in cas we ware gon longer than we calkalated. The furs ware in a good lodg that he mite keep himself warm. We left him in that sittuation and got back to the house whare we had left the goods by crossing the plaines. I found all safe and the clark had colected a leattel provishon but the provishons could not be sent to the boy on acount of the wather seating in so bad that the men would not under take to go across the plane. Sum days after it grew more modrat and thay seat of five in number and reacht him in fifteen days from the time we left him. Thay found him well but feeble. Thay gave him to eat moderately at first and he ganed strength. Thay went to work and put the furs on a scaffle out of the way of woods (wolves) or eney varment and all seat of for home. The day befour thay arived thay ware overtaken by a snow storm on the planes and could not sea thare way near night. Thay seat down on the plane thare beaing no wood nigh and leat the snow cover them over. Thay had thare blankets about them. In the morning—it was clear with ye wind norwest and freasing hard. Thay dug out of the snow and beaing weat in sum of thare feet thy was badley frosted tho not more than ten miles to walk. The boy ascaped as well as eney of them—I beleve the best. I had a long job to heal them but without the loss of a limb".

During the winter as in the previous year the Indians "found out whare we ware and came in with meet and furs to trade". With the approach of spring arrangements were made to bring the furs down from the cache to the post. "In the Spring I sent my people after the furs thay had put on a scaffel in the winter. Thay had an Indan hunter with them who kild them sum buffeloes. The men cut down small saplens and made the frames of two boates—sowed the skins togather and made bottoms to thare frames—rub'd them over with tallow which made tite anuf to bring the furs down to me whare I had canoes to receve them".

"The Spring is now advancing fast. The chefes cuming with a number of the natives to go with me to Mackenac to sea and hear what thare farther had to say". Pond was successful in his mission and "asembled eleven chefes who went with me besides a number of conscripted men. By the intarpretar I had the speach expland and the intenshun of the belts—and after we had got ready for saleing we all imbarkt and went down the River to its mouth".

Pond had left the St. Peter's river forever. Throughout his journal he constantly refers to the character of the country. "The intervale of the river St. Peter is exsalant and sum good timber—the intervels are high and the soile thin and lite. The river is destatute of fish but the wood and meadows abundans of annamels. Sum turkeas, buffeloes are verey plenty, the common dear are plentey, and larg, the read and moose deare are plentey hear, espesaley the former. I have seen fortey kild in one day by surrounding a drove on a low spot by the river side in the winter season. Raccoons are verey large. No snakes but small ones which are not pisenes. Wolves are plentey— thay follow the buffeloes and often destroy thare young and olde ones,—The natives near the mouth of the river rase plenty of corn for thare one concumtion."

Arriving at the junction of St. Peter's River with the Mississippi "we found sum traders who com from near the head of the Misseppey with sum Chippewa chefes with them. I was much surprised to sea them so ventersum among the people I had with me for the blod was scairs cold—the wound was yet fresh—but while we stade thare a young smart looking chef continued singing the death song, as if he dispised thare threats or torments. After we had made a short stay hear we imbarkt for the Planes of the Dogs" (Prairie du Chien).

Arriving at Prairie du Chien "we joined a vast number of people of all descripsions wateing for me to cum down and go to Macanac to counsel for these people had never bin thare or out of thare countrey except on a war party. It excited the cureosatay of everay nation south of the Lake of the woods and from that (place) was a number chefes which was more than two thousand miles. Indead the matter was intresting all parties espechaley to the trading party for the following reson—each of these nations are as much larger than eney of thare nighbering nations as the Inhabitans of a sittey are to a villeag and when thay are at varans property is not safe even traveling threw thare countrey". It was imperative for the safety of the trade on the Mississippi, which was of the greatest importance, that the difficulties should be settled.

PETER POND: FUR TRADER & ADVENTURER

Leaving Prairie du Chien "everay canoe made the best of thare way up (the Wisconsin) to the Portage and got over as fast as thay could". Pond apparently had special obligations to the Indians from St. Peter's river. "While we ware on the Portage one of my men informed me that thare was an Indan from St. Peter's River that was in Morneing for his departed friend and wished me to take of the morneing for he had worn it long anuf. I desired he mite cum to me which was dun. He was blacked with cole from the fire—hand and face. His haire was hanging over his eyes. I askd what I should due for him. He desird that his haire mite be pluckd out to the crown of his head, his face and hands washed and a white shirt put on him. I complied with the request and seat him on the ground—seat a cupel of men to work and with the asistans of a leattel asheis to prevent thare fingers slipping thay soon had his head as smooth as a bottle. He washt up and I put a shirt on him which made the fellow so thankfull to think that he could apear in a deasant manner that he could scairs contain himself. We desended the fox river to the botam of Greane Bay so cald and thare joind the hole of ye canoes bound to Macenac. The way ther was fair and plesant we all proseaded together across Lake Misheagan at the end of two days we all apeard on the lake about five miles from Macenac and aproacht in order. We had flags on the masts of our canoes—eavery chefe his flock. My canoes beaing the largest in that part of the cuntrey and haveing a large Youon (Union) flage I histed it and when within a mile and a half I took ye lead and the Indans followed close behind. The flag in the fort was histed—ye cannon of the garreson began to play smartley—the shores was lind with people of all sorts who seat up such a crey and hooping which seat the tribes in the fleat a going to that degrea that you could not hear a parson speak. At lengh we reacht ye shore and the cannon seasd. I then toock my partey to the commander (Capt. de Peyster) who treated us verey well. I seat with them an our and related the afare and what I had dun and what past dureing the winter. After interreduseing the chefe I went to my one house where I found a number of old frends with whom I spent the remainder of the day".

Pond's work on the Mississippi as a peace maker was over but the final arrangements had still to be made. "The people from Lake Supereor had arivd befour us and that day and next day the grand counsel was held before comander in the grate chamber befour a vast number of spectators whare the artickels of pece ware concluded and grate promises ware mad on both sides for abideing and adhearing closely to the artickels to prevent further blodshed the prinsapel of which was that the Nottaweses should not cross the Missacepey to the East side to hunt on thare nighbers ground—to hunt nor bread eney distarbans on the Chipewan ground. Thay should live by the side of each other as frinds and nighbers. The Chipewase likewise promis on thare part strickly to obsareve the same reagulations on thare part toward ye Nottawasis—that thay will not cross the river to hunt on the west side—after all the artickels ware drown up thay all sind them. The Commander then made a presant of a cag of rum to each nation and thay left the fort and went to thare camp whare thay seat round and ingoied thare present—sung a fue songs and went to rest in a verey sivel manner. The next day thare was a larg fat ox kild and coked by the solgers. All of the nations were biden to the feast. Thay dined to geather in harmoney and finished the day in drinking moderately, smokeing to gather, singing and britening the chane of frindship in a veray deasant way. This was kept up for four days when the offeser mad them each a present and thay all imbark for thare one part of thare cuntrey".

With the return of Peter Pond from the second and last venture in the Mississippi country, this journal comes to an end. He was at this time a trader and an explorer and thoroughly experienced in dealing with the Indians.

IV. THE NORTHWEST COMPANY

A. THE EXPLORER (1775-1778)

In the summer of 1775 after his arrival at Mackinac Pond decided to leave the Mississippi and to venture to Lake Superior, Grand Portage and the Northwest. It is probable that a number of considerations led him to take this step. He had doubtless paid the remainder of his debt to Mr. Graham,[5] and was now independent, able to purchase his own cargo, and to go where trade seemed most profitable. The furs on the Mississippi were becoming less important especially with competition and trade with the Plains Indians brought small supplies of such fine furs as otter and beaver. Proof of the finer quality of furs obtained in the north was shown in the profitable character of ventures which were being made to the Northwest. Finlay had wintered on the Saskatchewan in 1768. Three years later in 1771 Thomas Curry had wintered at Cedar Lake and taken down a large number of furs in 1772. Trade to the Northwest was profitable and competition with the

---

[5] A document dated 1778 protested on behalf of three houses, Alex. Ellice and Co., Felix Graham, and Todd and McGill against the unsatisfactory conditions of the Niagara Portage under the control of Mr. Stedman, and it is possible that Mr. Graham may have joined the exodus from Albany and New York to Montreal. He had apparently been engaged in the trade for two previous years 1776 and 1777 and it is possible that his connection with Pond continued. On the other hand the licenses give no reference to Mr. Graham and probably Pond sought new alliances. There is evidence to suggest that Pond was supplied by George McBeath and in turn by Simon McTavish in 1775 and that he may have been persuaded by these men to undertake at this date the voyage to the Northwest.

Hudson's Bay Company could be carried out with great effectiveness. Alexander Henry as well as Peter Pond decided in the same year that he could greatly increase his returns by going to the Northwest. It is possible that Pond decided that the peace between the Sioux and the Ojibway was not of a stable character and with Henry he may have heard of the rumblings of war in the colonies which promised further disturbances. Moreover competition was keen on the Mississippi and he had found it necessary to go far up the St. Peter to find new tribes of Indians. Pond had many friends among the traders and these in turn may have prevailed upon him to try the Northwest. Indeed it would not be difficult to accept a theory that Pond, Alexander Henry and Cadotte, the Frobishers, James Finlay, McGill and Patterson, and Holmes and Grant had decided upon the formation of an association for carrying on trade to the Saskatchewan in the spring of 1775. Probably the combination of considerations led him to decide for new territory; and more important than all was Pond's inherent restlessness which urged him to see new country. From the Mississippi it was customary to spend at least two months at Mackinac and this to Pond was wasted time. The Mississippi had little more to offer the trader.

In any case he bought a new cargo at Mackinac, hired his men and started for Sault Ste. Marie, the north shore of Lake Superior and Grand Portage. Taking his goods across the portage he followed the well known route to Lake Winnipeg; and Alexander Henry on August 18 shortly after he had left the Cree village at the mouth of the Winnipeg river notes that he was joined "by Mr. Pond, a trader of some celebrity in the northwest". Alexander Henry and Cadotte left the Sault with a large outfit on June 10, 1775. It is probable that Pond with a smaller outfit left much later and travelled much more rapidly. Together they proceeded along the east side of Lake Winnipeg. The following day the party was overtaken by a severe gale and forced to take refuge on "the island called the Buffalo's Head" but only after they had lost a canoe and four men. As far as possible they secured their provisions in the country and Henry writes that they took catfish of six pounds weight. On the twenty-first they crossed over to the south shore to Oak point where they spent some time in fishing. "The pelicans which we every where saw appeared to be impatient of the long stay". On September first they passed Jack-head river and on the seventh of September were overtaken by Messrs Joseph and Thomas Frobisher and Mr Patterson increasing the size of the party to thirty canoes and a hundred and thirty men. Henry writes "On the twentieth we crossed the bay together—on the twenty-first it blew hard and snow began to fall. The storm continued till the twenty-fifth by which time the small lakes were frozen over, and two feet of snow lay on level ground in the woods. This early severity of the season filled us with serious alarms for the country was uninhabited for two hundred miles on every side of us and if detained by winter, our destruction was certain. In this state of peril, we continued our voyage day and night. The fears of our men were a sufficient motive for their exertions." They reached the mouth of the Saskatchewan on the first of October.[6] Two miles up the river they came to the Grand Rapids "up which the canoes are dragged with ropes. At the end of this is a carrying-place of two miles". They reached Cedar Lake on October third and crossed that lake on the fourth. The following day was apparently spent in catching sturgeon using a seine of which the meshes were large enough to admit the fish's head and which was fastened to two canoes. On the sixth they started up the Saskatchewan. Two days later they began the "voyage before daylight making all speed to reach a fishing place, since winter was very fast approaching. Meeting two canoes of Indians we engaged them to accompany us as hunters. The number of ducks and geese which they killed was absolutely prodigious". At the Pas they found an Indian village of thirty families with a chief named Chatique, who succeeded in forcing them to give presents to the extent of "three casks of gunpowder, four bags of shot and ball, two bales of tobacco, three kegs of rum and three guns together with knives, flints, and some smaller articles". Moreover he followed them up and levied another keg of rum. "Every day we were on the water before dawn and paddled along till dark. The nights were frosty and no provisions excepting a few wild fowl, were to be procured. We were in daily fear that our progress would be arrested by the ice." On the twenty-sixth "we reached Cumberland House" which had been built the previous year for the Hudson's Bay Company by Samuel Hearne and which was now under the charge of Mathew Cocking, "by whom though unwelcome guests, we were treated with much civility".

    At Cumberland House the party broke up and Pond went to Fort Dauphin on the northwest corner of Lake Dauphin. He turned his canoes down stream to Cedar Lake, crossed over a portage to Lake Winnipegosis and went along its full length to Mossy river and to Lake Dauphin. Alexander Henry reached Beaver Lake on November

---

[6] Henry writes elsewhere that he was twenty-eight days crossing the lake.

first and the following day it had frozen over. Pond was doubtless forced to travel constantly to reach Lake Dauphin before the Mossy river had frozen. Unfortunately it is only known that Pond wintered on Lake Dauphin. As to why he should have gone to this district, as to whether he evaded Chatique successfully on his way down, and as to the time he arrived, conjectures can alone be made. Bain in his edition of the *Travels and Adventures of Alexander Henry* (p. 266) suggests that the Canadian traders stationed themselves on the three lines of communication, from the north, east and south, forestalling the Hudson's Bay Company at Cumberland House by meeting the Indians on their way to trade. Alexander Henry and the Frobishers wintered in the north, Peter Pond in the south and the remainder to the west on the Saskatchewan. Henry refers to his stock with that of the Frobishers and to the amalgamation of four interests on the Saskatchewan in 1775 but it is doubtful whether Pond was a party to the arrangements. He was probably an independent trader with two canoes and the probable complement of four men each. The other interests represented the great bulk of the canoes and they were in a position to take first choice and to control the best fields north to the Churchill and along the Saskatchewan. In the fleet of thirty canoes, Pond had two, Henry and Cadotte eight, the Frobishers six, and Patterson the remaining fourteen for Fort des Prairies. The twelve canoes of the Frobisher license in Montreal became twenty small canoes in the Northwest. To Pond with his smaller outfit, if he had not already become a member of the Saskatchewan group, was left the alternative of joining, or of co-operating with them by seeking a strategic wintering place. To the south on Lake Dauphin he was in a position to secure a supply of food from the buffalo as he was on the edge of the plains and to trade with the Assiniboines with whose language and customs he had become acquainted among the Yanktons on the Mississippi. Moreover as Bain suggests it was a point at which he could trade with the Assiniboines and other Indians going to Cumberland House or to Hudson Bay. It appeared to Pond or to the group as the most promising locality for a small outfit, but we know nothing about this venture other than that the letters P P 1775-6 are marked on Lake Dauphin on Pond's map of 1785.

Pond probably returned to Grand Portage or to Michilimackinac to dispose of his furs and to acquire another outfit. He was within striking distance of his old base at Michilimackinac and it is possible that he returned to that point. This may have been a deciding factor in his determination to winter at Lake Dauphin. In any case he would have learned as Alexander Henry did of the troubles which had broken out in the colonies and of the capitulation of Montreal and he would have found that the old source of supplies at New York and Albany was effectively cut off.

Again we are forced to rely upon meagre information as to his second venture. The map of 1785 shows the letters P P 1776-7 on a post at the junction of the Sturgeon river and the North Saskatchewan a short distance above the forks. The rough map copied by Ezra Stiles shows at this point the inscription *Capt. Pond wintered 2 y with 160 men*. Other inaccuracies on this map lead one to conjecture that he wintered at this point in 1776-7 and 1777-8 with 4 canoes and 16 men. Pond, or the group, had decided that Fort Dauphin was not a profitable post at least for four canoes or a larger outfit and that the Saskatchewan offered the largest returns. This decision implied probable rearrangements in his plans. The longer upstream journey on the Saskatchewan almost certainly meant a shift from the base at Michilimackinac to Grand Portage. This would necessitate reliance on some connection at the latter point which would bring up an outfit from Montreal to Grand Portage and take down his furs to be sent to England. He would find a partnership or an arrangement with Montreal firms essential after the beginning of war with the colonies.

Again we can only conjecture as to the arrangements. Pond as has been suggested may have proceeded to the Northwest on his first voyage in 1775 with the support of George McBeath and Simon McTavish. The latter was engaged in forwarding rum from Albany to Detroit in 1774, and in 1775 he owned in partnership with George McBeath a boat of thirty tons on Lakes Erie and Huron. He apparently left the Albany trade after the outbreak in the colonies and made his headquarters at Montreal. With Bannerman he owned a small boat (perriauger) on Lake Superior which wintered at Grand Portage in 1775-6. Pond's supplies may have been taken up on this boat on his first venture. In 1776 McTavish took to England about £15,000 worth of furs presumably from Grand Portage and possibly including Pond's returns. McBeath however appears to have acted as agent for the house with which Alexander Ellice was connected. The firm of Phyn and Ellice had been forced to leave Albany and some members of the firm moved to Montreal. James Ellice died in Montreal in October 1787. James Phyn merchant on Old Broad Street, London was a correspondent of Alexander Henry's in 1781. In any case George McBeath took out a license for the first time in 1777 to take five canoes to Grand Portage and Alexander Ellice acted as guarantor. It is probable that Pond knew George McBeath since the latter had been at Michilimackinac at least as early as 1768. It is also probable that Pond was not averse to forming a connection with the old Albany firm of Phyn and Ellice with the new Montreal headquarters. The conjecture is submitted that in 1776 Pond at Michilimackinac arranged to have the next season's goods brought to Grand Portage and that the license secured by George McBeath in 1777 included goods which were for his outfit. Pond and McBeath later became partners who were supplied by Phyn and Ellice and it is quite possible that the arrangements began at this early date. With these plans Pond proceeded to the post on the North Saskatchewan after his trip down from Fort Dauphin confident that in the following year McBeath would meet him at Grand Portage with a new outfit and prepared to take his furs. And according to expectations Pond probably met McBeath in 1777 at Grand Portage and returned for his second winter at the post on the North Saskatchewan.

It is impossible to state how far Pond was still an independent trader and as to whether the arrangements with other traders on the Saskatchewan became more definite and tangible. But Pond had apparently established his post above the other Canadian posts and if he was still an independent trader, as seems to be the case, he was acting with his usual superb skill as a trader in reaching the Indians farthest up the Saskatchewan. In the same year (1776-7) that Pond had decided to winter on the North Saskatchewan in the hope of securing a larger return than he had secured at Fort Dauphin, the Hudson's Bay Company had also decided on a similar move. Cumberland House had not been adequate to check the Canadians. As Bain has shown the Canadian traders had made a successful countermove by forestalling the Indians in the more important routes. As a further move in the struggle the Hudson's Bay Company decided to establish a post farther up the Saskatchewan than that of any of the Canadian posts. Consequently while Pond was establishing his post above the Forks the Hudson's Bay Company under Turnor's direction was building a post still farther up the North Saskatchewan at Hudson's House not far from the present site of Prince Albert. Pond was doubtless successful the first year 1776-7 since both posts were newly established, but in the second year 1777-8 competition apparently led him to think of fresh fields of larger profits. He may have found himself in possession of a large stock of goods at the end of the winter and a relatively small supply of furs.

An alternative field was suggested in the successful expeditions which had been made by Henry and the Frobishers on the Churchill in 1775-1776 and of Thomas Frobisher who had wintered at Isle à la Crosse Lake in 1776-7. The best furs were to be obtained in the north and the profit from the trade with the Indians on the Churchill River were large. The Hudson's Bay Company had advantages on the Saskatchewan with direct access to the Bay but on the Churchill where the skill of canoe-men was at a premium it was difficult to follow. The Canadians had a more promising field. It was this area which Pond had decided to attack.

On the other hand the area had decided difficulties. From the standpoint of trade the canoe had definite limits. These limits were primarily those of food supply. The canoe could not carry a large quantity of food in addition to the trading goods and men. To penetrate to the Churchill river and beyond meant dependence on the food supply of the country and especially on fish. It was extremely difficult with the relatively large complement of men for a canoe to travel rapidly and to secure an adequate supply of fish. The experience of Pond, Henry and the Frobishers on Lake Winnipeg in 1775 had shown this very clearly. Incidental to the problem of an adequate food supply was the shortness of the season in which to make the journey from Grand Portage to the interior. Alexander Henry and his party had secured liberal supplies of rice on the route to Lake Winnipeg and depended on fish for the remainder of the journey. Thomas Frobisher had solved the problem in part by continuing from the fort built on the Churchill River in 1776 with the goods left over from the trading of Henry and the Frobishers and wintering at Isle à la Crosse lake where an adequate supply of fish was available. Although Frobisher had planned to return with the Indians to Lake Athabaska apparently he was unable to carry it out. If in spite of the help of the Indians he had failed and had been obliged to stop at Isle à la Crosse lake how far was it to be expected that Pond would succeed.

The account of the expedition available in Alexander Mackenzie's *A general history of the fur trade* (London, 1801) is probably accurate since Mackenzie doubtless gained this information during the winter of 1777-8 with Pond in Athabaska. The paragraphs may be quoted in full. "The success of this gentleman (Thomas

Frobisher in 1777 at Isle à la Crosse) induced others to follow his example and in the spring of the year 1778 some of the traders on the Saskatchiwine river finding they had a quantity of goods to spare agreed to put them into a joint stock, and gave the charge and management of them to Peter Pond, who, in four canoes, was directed to enter the English river, so called by Mr. Frobisher, to follow his track, and proceed still further; if possible to Athabaska, a country hitherto unknown but from an Indian report. In this enterprise he at length succeeded, and pitched his tent on the banks of the Elk River by him erroneously called the Athabaska river about forty miles from the lake of the Hills into which it empties itself".

"Here he passed the winter of 1778-9; saw a vast concourse of the Knisteneaux and Chepewyan tribes, who used to carry their furs annually to Churchill the latter by the barren grounds, where they suffered innumerable hardships and were sometimes even starved to death. The former followed the course of the lakes and rivers, through a country that abounded in animals, and where there was plenty of fish; but though they did not suffer from want of food, the intolerable fatigue of such a journey could not be easily repaid to an Indian, they were therefore highly gratified by seeing people come to their country to relieve them from such long toilsome and dangerous journies; and were immediately reconciled to give an advanced price for the article necessary to their comfort and convenience. Mr. Pond's reception and success was accordingly beyond his expectation, and he procured twice as many furs as his canoes would carry. They also supplied him with as much provision as he required during his residence among them and sufficient for his homeward voyage. Such of the furs as he could not embark, he secured in one of his winter huts and they were found the following season, in the same state in which he left them." (pp. xii-xiii)

No direct record of his first expedition to Athabaska has been made available but a letter from Alexander Henry to Joseph Banks dated Montreal, Oct. 18, 1781 enclosing a memorandum on an overland route to the Pacific printed in L.J. Burpee, *Search for the Western Sea*, (Toronto, 1908) pp. 578-587, throws some light on the excursion. Henry does not mention his informant but undoubtedly it was Pond with whom he probably had numerous conversations in the winter of 1779-80 and he may have seen Pond in the summer of 1781 after his second visit to Athabaska. Henry had not penetrated beyond Isle à la Crosse lake and the account of the route beyond that point was almost certainly based on Pond's information. Henry recommends securing guides at Churchill river "as the natives inhabiting these parts, seems sepparated, and unconnected with the Southern Indians theire language differing but little from the kristinay the passages very difficult to find in the river on acco't of its breadth and number of islands—and great lakes, here the canoes must be refitted and when ready, proceed up this river, in the streights of which, are great rappids and dangerous places from the great body of watters, coming down from the melting of the snow in this season and proceed to Orrabuscaw carrying place, two hundred and fifty leagues, from where you fall on the Great River, course, near west, supposed to be in sixty degrees lattitude, day nineteen hours long the twenty second of June, the north side of this river appears a barren mountaneous country mostly bare rocks great quantity of fish in this river, the natives appear, as you proceed to the Westward, to increase in number". With this section Henry was familiar. Pond probably gave him a description of the portage which Henry or Pond locates at 60 N. Lat. 140 W. Long. "at this carrying place, provisions must be procured from the natives dry'd mouse and rein deer to subsist the people while carrying over this carrying place which is twelve miles long, and as far as Orabuscaw Lake, it will take six days to carry over, at the other end of which you fall on the river Kiutchinini which runs to the westward, takes its rise to the northward of this place, current, gentle, the land low and marshy great plenty of wild fowl in the fall—the season by this time will be advanced, it will be necessary to prepare for winter, build houses etc. the frost very severe, imploy the natives to hunt, for the subsistance of the men—which is mostly flesh, dry'd buffaloes meat, and mouse deer, it is not only the provisions for the winter season, but, for the course of next summer must be provided which is dry'd meat, pounded to a powder and mixed up with buffeloes greese, which preserves it in the warm seasons here every information must be procured from the savages, relating to the course of this river, the inhabitants of the lower parts the Orabuscaws, makes war on the Kitchininie therefore, it will be necessary to procure a peace between the two nations which would be no difficult matter for the natives, from all parts, hearing of Europeans being here, will come in the winter season to bring provisions on acco't of European trinkets, by which means, intelligence may be procured, and conveyed to any distant parts, it is common for these wandering tribes to remove two hundred leagues in the winter season, and carry their tents, family's and everything belong to them, when every thing is ready and provisions procured for the summer, as no dependance can be put, on what you are to receive in an unknown part. Every precaution should be taken, but very often, must expect to be dessappointed. A new sett of interpretors and guides must be procured, which can always be had, from the difft nations and proceed down this great river, untill, you come to the sea, which cant be any very great distance. Suppose it should be thirty, or forty degrees of longitude unless some accidents should entervene, it can be done in thirty days which will be in July[7], here an establishment may be made in some convenient bay or harbour, where, shipping, may come to. In the meantime a small vessel may be built, for coasting and exploaring the coast which can be no great distance from the streight which seperates the two contenants". Henry continued the memorandum showing the advantages of an establishment to exploration and trade, the probable cost of the expedition and the distances to be covered. Based as this memorandum must have been on Pond's observations it illustrates clearly the organization of the food supply in the Athabaska district and the conception which had been gained at this date of the possibilities of what is at present the Western Arctic department of the Hudson's Bay Company.

Pond had succeeded in overcoming the difficulties by proceeding inward from Cumberland House with the surplus goods of the winter trade early in the season and not delaying by a trip to Grand Portage, and by securing an ample supply of provisions at his post on the Athabaska river. It was a remarkable tribute to Pond's ability that he should have been chosen by the other interests to represent them and that he should have been entrusted with four canoes of goods. He undoubtedly impressed his colleagues as a man of sterling character, of courage, aggressiveness and ability. To Pond must be given the credit of being the first white man to cross the Portage la Loche and of discovering the Athabaska river and Lake Athabaska. He had succeeded where Thomas Frobisher had failed. This success was in part the result of organization. He was probably supplied with dried meat as a supply of food and which in conjunction with the excellent fishing of the Churchill River waters and the short distance from Cumberland House made it possible to reach the district whereas Frobisher was dependent on fish. After reaching the Athabaska he became dependent on the pemmican and meat supplied by the Indians of that area. Pond

---

[7] Alexander Mackenzie began the return journey from the mouth of the Mackenzie on July 16.

was an explorer as well as a trader and as he had been first on St Peters River, had gone farther than the Canadian traders in the North Saskatchewan, he was first into Athabaska. He traded successfully with large numbers of new tribes of Indians as in the days of 1774-5 on the Mississippi.

Success in the Athabaska district required an organization in the neighbourhood of Cumberland House and an organization of provision supplies in the Athabaska district for the return journey. It involved further the development of provision supplies in the Red River district adjacent to the southern end of Lake Winnipeg. It was not possible to cover the long distances and depend on the supplies obtained along the way. The use of pemmican[8] which may have been a contribution of the Athabaska Indians facilitated the development of this organization. Later improvements followed with the establishment of an advanced depot at Lac la Pluie. This wide organization necessitated extensive co-operation between a variety of interests located in separate territories. The Athabaska district was important to the development of the Northwest Company and the work of Peter Pond was of fundamental importance.

B. THE ORGANIZER (1778-1790)

Pond had been unusually successful in his trip to Athabaska in 1778. He came out the following summer (1779) and went down to Montreal. He had left Montreal in 1773—six years of expeditions to the Mississippi and to Athabaska—and in 1779 he was thirty-nine years of age. In the fur trade of the Northwest he was an old man. It was primarily conducted by young men. Alexander Mackenzie left the

---

[8] Alexander Henry mentions dried meat but not pemmican in his journal. He describes pemmican in detail in his letter of 1781 which was dependent on Pond's observations. "It is not only the provisions for the winter season, but, for the course of next summer, must be provided which is dry'd meat, pounded to a powder and mixed up with Buffeloes greese which preserves it in the warm seasons here." It is quite possible that Pond solved the problem of the Athabaska trade by discovering the use of pemmican rather than dried meat. Alexander Mackenzie refers especially to the Chipewyans in his description of the method of making pemmican.

Northwest at the age of 30, David Thompson left at 40, and Alexander Henry, Pond's contemporary at 37. It was a hard life which demanded much and the profitable character of the trade in the early period enabled men to retire at an early age and at the same time gave the organization the full advantage of youth and energy. According to the standards of the trade Pond should have left the Northwest after he had discovered Athabaska in 1779. But for him the mystery of the Mackenzie river drainage basin had not been solved.

When he came out in 1779 presumably the arrangement with his supporters did not end as there remained in Athabaska a large supply of his furs in cache. This arrangement was apparently continued through the appointment of Waden to take his place during his absence to Montreal. According to Mackenzie, Waden wintered in 1779 at Lac La Ronge. The agreement of 1779 which provided for his selection was a formal arrangement in which each of the traders was given a definite share. McBeath and Co. which probably included Peter Pond held two shares. In 1780 McBeath's name does not appear among the grantees since Pond was in Montreal. In that year the latter held two licenses, nos. 11 and 19, each permitting to take to Grand Portage 2 canoes 250 gallons of rum, 20 rifles, 600 lbs. gunpowder, 8 cwt. ball and shot, and valued at £750. With these goods he probably went in to Athabaska wintering there again in 1780-1 and bringing out the furs which had been cached in 1779. He probably came out to Grand Portage in 1781 and secured the goods which included 4 canoes, with 33 men, 600 gallons of rum, 50 gallons wine, 50 rifles, 600 lbs. powder and 20 cwt. of shot valued at £3000 which had been brought up under a license granted to McBeath, Pond, and Graves with McBeath and R. Ellice as guarantors. In the same year Mr. Waden's "partners and others engaged in an opposite interest, when at the Grand Portage agreed to send a quantity of their goods on their joint account, which was accepted and Mr. Pond was proposed by them to be their representative to act in conjunction with Mr. Waden".

The quotation from Sir Alexander Mackenzie's *voyages* is confused as to dates. According to his statement Waden was killed in 1781 but according to the sworn statement of "Joseph Fagniant de Berthier ordinairement voyageur dans le Pays en hault" he died in the month of March 1782 "dans le pays en hault du lac de la Riviere aux Rapids dans la Riviere des Anglois—dans un petit Fort avec Peter Pond et Jean Etienne Waden commercants". According to Mackenzie "Mr. Waden, a Swiss gentleman, of strict probity and known sobriety, had gone there (Lac la Ronge) the year 1779 and remained during the summer of 1780". In this year Pond is alleged to have wintered with Waden, the latter meeting his death about the end of the year 1780 or the beginning of 1781.

Waden's position in the fur trade of the Northwest warrants consideration. The licenses issued in his name throw some light on his importance. In 1772 a license permitted him to take a canoe and 8 men to Grand Portage giving his own name as security; in 1773, 2 canoes with 16 men; in 1775, 2 canoes with 15 men, Mme Waden, as guarantor; in 1777, 3 canoes and 17 men J.E. Waden as guarantor; in 1778, 3 canoes and 23 men with Richard Dobie and J. McKindlay as guarantors. In 1779 his name does not appear but that of V. St. Germain appears in two licenses both having J. McKindlay as guarantor and one license reading for 1 canoe and 10 men the other for 2 canoes and 18 men. The general agreement of that year in which Waden and Co. held two shares and which was represented by Waden at Lac la Ronge found the firm represented in Montreal by St. Germain. In 1780 the firm of Waden and St. Germain secured two licenses one for 4 canoes and the other for one canoe. In 1781 the firm of Waden and St. Germain with St. Germain and Dobie as guarantors secured a license for 4 canoes and 38 men.

The problems of the fur trade from 1778 to 1783 are extremely complex.[9] The sources of information are meagre and in many cases biassed. Conjectures based on this information will be subject to revision but an attempt to reconstruct the history may be made. The evidence points very directly to the formation of a large organization probably as early as 1775. This organization was apparently built directly around the Frobishers, the McGills, the Ellices, and McTavish or the larger houses of Montreal. In the agreement of 1779 this general vague organization possibly included Todd and McGill, B. and J. Frobisher, McGill and Patterson, McTavish and Co. and McBeath and Co. a total of ten shares of the sixteen. Pond was probably despatched by these interests to Athabaska in 1778, the two larger houses that of McTavish and the Frobishers joining to support the venture. Pond was well known to Todd and McGill as he came up from Montreal to Michilimackinac with them in 1773 and to the other traders including the Frobishers and Henry and he was trusted as a thoroughly competent

---

[9] See H. A. Innis, *The Northwest Company*, Canadian Historical Review, Dec., 1927.

trader. McBeath, and the firm of Phyn and Ellice were probably his direct supporters and in turn he would be connected with McTavish. He had gone in to the Northwest in 1775 with the Frobishers and Alexander Henry and was the first to penetrate the Athabaska department in 1778. He was the logical choice of these interests to push the trade into Athabaska. The difficulties of the fur trade were the result of the struggle which developed between this large group and the number of small traders who had no such allegiance. In the small group Waden, later Waden and St. Germain, was a typical representative, and also Oakes and Co. a firm which had been supported by Lawrence Ermatinger; Ross and Co. and probably Holmes and Grant, although this firm had early affiliations with the larger group. It was possibly as a concession to this group that Waden "of strict probity and known sobriety" was chosen to take Pond's place in the trade with the Athabaska Indians in 1779. Pond as already shown probably returned to Athabaska in 1780 and came out to Grand Portage in 1781. Waden possibly came out to Grand Portage in 1780 but in view of Pond's visit it would not be necessary for him to come out in 1781 and he is reported as summering in the district.

In 1781 the problem arose as to the appointment of representatives which would satisfy all the interests. Waden was in the interior and the small traders were probably insistent that he should remain as their representative. But Pond had been the representative of the larger group. He was familiar with the trade and the Indians of Athabaska and was, through his long experience as a trader, very successful. A compromise was apparently reached in which Pond was chosen to represent the larger interests and to trade a joint stock with Waden at Lac la Ronge. It is not difficult to imagine the causes of contention which would arise between Pond and Waden, the one chosen to carry out the policies of the larger group and anxious to develop the Athabaska department, the other interested in the policies of the smaller group anxious to keep down expenses and content to wait at Lac la Ronge for the Indians to bring out the furs from Athabaska in the spring. Pond was selected by the larger group because its members were confident in his ability to carry out their policy and Waden was probably chosen by the members of the smaller group for a similar reason. Moreover Alexander Mackenzie wrote "two men of more opposite characters could not perhaps have been found. In short from various causes their situations became very uncomfortable to each other and mutual ill-will was the natural consequence". The sworn testimony of Joseph Faignant de Berthier is probably all the evidence that is available as to these natural consequences. "Au commencement du dit mois de mars (after a winter living in close proximity in a small fort) vers le neuf heures du soir, le deposant s'etant retiré dans sa maison qui etoit à côte et touché la maison du dit sieur Waden, de chez qui le deposant revenoit et etoit après se dechosé le deposant en dix minutes ou environ après avoir arrivé, ayant quitté le dit Waden sur son lit avoit entendu tirer deux coups du Fuzils, l'un après l'autre subitement dans la maison du dit Waden, sur quel le deposant envoyoit un homme pour voir ce que c'etoit qui alloit et revenoit et rapportoit a ces deposant que Mons. Waden etoit a terre ayant recu un coup du Fuzil sur quel le deposant se levat et courut immediatement au dit Waden et lui trouvat a terre a coté de son lit ou le deposant lui a quitté le peu de tems au paravant, et trouvat son jambe gauche cassé de son genu, jusqu'en bas; qu'en aprochant le dit Waden lui disoit. *Ah mon amis je suis mort* sur quel le deposant a assayé de dechirer ses culots a metapes pour l'examiner, et trouvat la marque de poudre sur son genu, et de trois ou deux balles avoitent entré et trouvat la jambe casse jusqu'en bas ou les deux balles ont sortis a derier, les ayant trouvé sur la place; que le dit sieur Waden disoit au deposant de trouver la Boam Turleton et d'arrêter le sang; qu'ayant lui demandé ce qui lui a fait cela; il a repondu je vous le dirai; mais ayant perdu bien du sang alors, il n'etoit pas capable de lui dire plus; Que le deposant en entrant chez le dit Waden sur cette occasion a percu le dit Peter Pond et Tousaint Sieur, sortant de chez le dit Waden et entrant chez eux et trouvat, un fuzil vide et un autre cassé dans la ditte maison de Waden, et apercu que cel qui etoit vide etoit dernierement tiré mais que l'autre etoit emporté. Que le deposant en entrant chez M. Waden apres le coup a vu le dit Peter Pond et Toussaint Sieur á la porte et le dit Sieur demanda du dit Waden si c'etoit lui le Sieur qui l'avoit tire. Que le dit Waden a reponder *allez vous en tous les deux que je ne vous voyez plus*. Que la dessus deux nommes amenerent le dit Tousaint Sieur pour coucher et Peter Pond entra chez lui, qu'environ un mois devant le dit Peter Pond et le dit Waden se batterent ensemble et le meme soir que le dit Waden a ét tué. Environ une heure devant soupé Peter Pond se quarrelloit et disputoit avec le dit Waden. Que le deposant a grande raison a croire que c'etoitent les dits Peter Pond et Toussaint Sieur ou un de eux qui ont tué le Sieur Waden et pour le present n'a rien plus a dire. Affirmé pardevant moi a Montreal le 19 May 1783 (signe) Neveu Sevestre C.C. (a true copy of the original). This differs from Mackenzie's account which states "Mr. Waden had received Mr. Pond and one of his clerks to dinner; and in the course of the night, the former was shot through the lower part of the thigh, when it was said that he expired from the loss of blood, and was buried next morning at eight o'clock". And this in the hard frozen ground of Northern Saskatchewan in March! Alexander Mackenzie was not partial to Pond but he states that "Mr. Pond and the clerk were tried for this murder at Montreal and acquitted". No record has been found of the trial at Montreal but presumably all the available evidence was taken including the above testimony and Pond was acquitted. It is not customary to go behind the verdict of the courts but Mackenzie in his work published in 1801 probably knowing that Pond was at that time impoverished and that he had severed his connection with the Northwest Company and Canada, apparently felt that he was safe in writing "nevertheless their innocence was not so apparent as to extinguish the original suspicion". This was doubtless the view taken by the small group of traders whose interests Mackenzie had always represented but the verdict of the courts cannot be disregarded.

Despite the unsatisfactory character of Mackenzie's account it is the only one available. It is necessary as far as possible to make corrections for the bias and to assume that the remainder is accurate. According to this account Pond despatched Toussaint Sieur (the above mentioned clerk) "to meet the Indians from the Northward who used to go annually to Hudson's Bay; when he easily persuaded them to trade with him, and return back that they might not take the contagion which had depopulated the country to the eastward of them; but most unfortunately they caught it here and carried it with them, to the destruction of themselves and the neighbouring tribes". It is not possible to determine whether Pond came down to Grand Portage in 1782 or whether his clerk was sent but it is probable that Pond went down. The news of Waden's death doubtless precipitated a disruption of the agreement and again divided the trade between the small group and the large group. Both groups therefore "began seriously to think of making permanent establishments on the Missinippi river and at Athabaska for which purpose in 1781-2 (probably 1782-3) they selected their best canoe-men being ignorant that the small-pox penetrated that way. The most expeditious party got only in time to the Portage la Loche, or Methy-Ouinigam, which divides the waters of the Missinippi from those that fall into the Elk river, to despatch one canoe strong-handed, and light loaded to that country; but on their arrival there, they found, in every direction, the ravages of the small-pox, so that from the great diminution of the natives, they returned in the spring with no more than seven packages of beaver. The strong woods and mountainous countries afforded a refuge to those who fled from the contagion of the plains; but they were so alarmed at the surrounding destruction, that they avoided the traders and were dispirited from hunting, except for their subsistence". No intimation is given in this account as to whether Pond was in command of the party to Athabaska but there is reason to believe that the traders successful in reaching Athabaska belonged to the small group. Their lack of knowledge of the country as well as the desolation of the small pox would account for the small returns in furs. Moreover Pond's map of 1785 shows that he wintered on Isle à la Crosse lake in 1783. He probably represented the interests of the larger group and was unsuccessful in getting into the Athabaska. His trip to Grand Portage and return would place him at a disadvantage with the other interests who were able to start off directly news was received of Waden's death. It is not known whether he went down to Grand Portage in 1783 but the scarcity of furs may have warranted the abandonment of the trip and let him proceed to Athabaska from Isle à la Crosse lake. Madame Waden had asked for his apprehension at Grand Portage in 1783 but he did not come out to Montreal until 1784. Probably he went direct from Isle à la Crosse lake to the Athabaska department in the spring of 1783.

In any case he went into Athabaska in 1783 and came down to Grand Portage and Montreal in 1784. The memoir which accompanies his map of 1785 is dated Arabosca 15 Mars 1784. From the map and the memoir it is possible to reconstruct in part the activities of the winter. If the foregoing analysis is correct this was Pond's third winter in Athabaska. In the two earlier visits he probably had little time for exploration because of the shortness of the season but if he went in early in 1783 from Isle à la Crosse lake he may have had time to carry out a more adequate survey. Certainly he gained from the Indians a vast amount of accurate information. He probably explored Lake Athabaska in part but it is extremely doubtful whether he went below it. Peace river is shown as running directly into Slave Lake and although he might have descended the Slave river without noticing that the Peace came in near Lake Athabaska, it is extremely doubtful. He did however from the Indians learn the approximate location of Slave Lake and of Bear Lake and the entrance of the river into the Arctic Ocean. The map is not accurate but from the standpoint of the fur trader it gives all the necessary information. He was fortunate in meeting Indians who had accompanied Hearne on his expedition to the mouth of the Coppermine and to elicit from them information gained on that visit. This information should be noted. "J'ai tenu conseil avec 40 des naturels qui vivent a une petite distance dela mer du N.O. Les autres tributs les appellent les gens du couteaux rouge. Ce nom leur vient de ce que presque tous leurs articles son faits de cuivre rouge dont leur pays est rempli. (no.6) Ils confirment le flux et reflux des eaux dans cette mer. Ils assurent qu'ils ne connoissent aucune terres nord, que les cotes courent vers l'ouest, que la navigation des rivieres qui tombent dans cette mer est ouverte dans le commencement de l'eté; qu'ils ont vu plusieurs foix des isles de glace, flottant sur ces (?) arages (no.7) Ice herring (Hearne) pendant les annees 1773, 1774, et 1775 a entrepris un long et penible voyage avec les naturels font aller examiner les grandes mines de cuivre. On a etouffé les plus petite circonstances de son voyage et des ses decouvertes. Jai recu ces details des sauvages qui l'accompagnerent et aujourd'hui il est defendu a touts personnes d'aller vers l'ouest".

During his winters in Athabaska and especially in 1783-4 he had come in contact with a large number of the Indians of the Mackenzie river district. In 1782 Fort Churchill was destroyed by the French and Matonnabee the trader of the northern Indians who had been influential in bringing trade to the Hudson's Bay Company committed suicide. The organization by which trade was carried on from the Mackenzie river drainage basin to Fort Churchill was destroyed. The establishment of Pond's fort in Athabaska was a final blow to Hudson's Bay Company supremacy and it finally brought into the hands of the Canadian traders the trade of the large area drained by the Mackenzie river and its tributaries. Mackenzie[10] writes "Till the year 1782, the people of Athabaska sent or carried their furs regularly to Fort Churchill, Hudson's Bay.—At present, however, this traffic is in a great measure discontinued".

In 1784 Pond came out to Grand Portage and was among those interests which arranged for the formation of the Northwest Company. The organization of the Athabaska district demanded strong capital support and it was felt that the agreement which began during the previous decade and was broken through the disruption of the small

---

[10] Hearne wrote from Churchill in 1780, "The Canadians have found means to intercept some of my best Northern Traders. However, I still live in hopes of getting a few [furs] from that quarter [Athabaska]". *David Thompson's Narrative, op. cit.* p. XXIII.

traders should be renewed and made more permanent. Arrangements were made for signing the agreement but the small traders continued to show their hostility to the large group and they were especially hostile to any agreement which threatened their interests. These small interests had suffered through the loss of Waden but they were as yet unwilling to surrender. The sixteen share agreement which was finally decided upon by the larger interests probably included 6 shares divided evenly between the two large houses of McTavish and B. and J. Frobisher, two shares each, held by Small and Montour probably the direct representatives of those houses in the interior, two shares each, to McBeath and Grant, and one share each, to Holmes and Pond. No provision was made for the small individual traders including Peter Pangman and John Ross. The latter had sent one canoe and 9 men to Grand Portage in 1779 and in the following year the firm of Ross and Pangman sent 4 canoes. Ross was given one share in the agreement of 1779 but no mention is made of these individuals in the licenses of later years. On the evidence given by the licenses it is difficult to understand the basis on which they could expect to be included in the agreement of 1783-4. They entered the trade at a very late date, were probably included in 1779 through the force of circumstances since in that year licenses were given out very late and closely restricted but certainly their contribution in capital or experience could not be regarded as important. Since they were excluded they decided to arrange for the formation of a new company.

At the time they were supported very curiously by Peter Pond. Pond was dissatisfied and with some reason because only one share had been given to him whereas McBeath his old partner secured two shares. Pond had been in the front line and had served with great energy and success for the larger group of traders in the Northwest and in Athabaska. He consequently expected that he should be rewarded for having borne the heat of the day. So far as is known McBeath, his partner, had never been beyond Grand Portage. But capital and not skill was the determining factor and Pond was assigned one share. He refused to sign and came down to Montreal at first apparently determined to throw in his lot with the small traders Ross and Pangman for whose defeat he had earlier been so largely responsible.

Little is known of his activities in 1784-5 during his year at Montreal. It may be assumed that he reported for trial for the murder of Waden and was acquitted. There can be little doubt that he reported to his old friends in Montreal his discoveries in Athabaska and that he soon forgot his grievances against the Company. It was hardly to be expected that he would continue with the small traders since he had nothing in common with them. Taking advantage of his loyalty to his old friends the Frobishers probably had little difficulty in persuading him to take his share in the Company. He may have visited his home at Milford as he apparently on the first of March 1785 presented a map to Congress. In the introduction to the note for which no date is given there is written. "Il est reparti pour constater quelques observations importantes". On April 18, 1785 at Quebec he signed a memorial probably written by the Frobishers. The memorial gives some hint as to the means by which he had been prevailed upon to rejoin his friends. He writes "Your memorialist begs leave to assure your Honour that the persons connected in the Northwest Company are able and willing to accomplish the important discoveries proposed in their memorial to His Excellency General Haldimand; provided they meet with due encouragement from government, having men among them who have already given proof of their genius and unwearied industry, in exploring those unknown regions as far as the longitude of 128 degrees west of London; as will appear by a map with remarks upon the country therein laid down, which your memorialist had lately the Honour of laying before you for the information of government, and the Company will procure at its own expense such asistants as may be found necessary to pursue the work already begun, until the whole extent of that unknown country between the latitudes of fifty-four and sixty-seven to the North Pacific ocean is thoroughly explored, and during the progress of this enterprize the Company will engage to transmit from time to time to His Majesty's Governor of this province for the information of government, correct maps of those countries and exact account of their nature and productions, with remarks upon everything else useful or curious that may be met with in the prosecution of this plan". Pond was apparently promised a free hand in carrying out his explorations of the Mackenzie river drainage basin and of rounding out the information which he already acquired.

  In 1785 Pond returned to Grand Portage and to Athabaska. But he was again to come in conflict with the small traders who were dissatisfied with the agreement of the Northwest Company. Ross and Pangman had been successful in enlisting the support of Gregory, McLeod and Co. of Montreal. These interests had been engaged with the Detroit and Michilimackinac trade but seeing the prospect of decline and loss of this trade through the success of the American

Revolution were ready to listen to proposals for entering the Northwest trade. It was not a promising venture. Of the members of the new concern John Gregory had never as far as is known engaged in the trade of the Northwest, Peter Pangman and John Ross had been in the trade for a short time but were apparently not very successful traders, Alexander Mackenzie who had been in the counting house of Mr. Gregory for five years and had never been beyond Detroit, Normand McLeod, was a dormant partner, Duncan Pollock "had served his time in the post office of Quebec but had lately been engaged in the trade among the Indians of Michilimackinac and of course was understood to be learned in Indian affairs," of whom (Roderic Mackenzie writes "His conduct was often very unpleasant to me and at length brought on an explanation which placed us on a good footing for the rest of the voyage if not orever after"), Laurent Leroux apparently knew something of the trade, James Finlay, was a son of James Finlay the early trader, and a brother-in-law of Gregory, and Roderic Mackenzie came to Canada in 1784. "The guides, commis, men, and interpreters were few in number and not of the first quality". These men were expected to compete with the traders of the Northwest Company and their success was not great.

John Ross probably accompanied by Laurent Leroux was despatched to Athabaska to compete with Peter Pond. Nothing is known of the movements of Pond or of the success of Ross during the winter of 1785-6. Pond's map of 1790 has remarks on it to the effect that he made excursions from his post on the Athabaska during the summer of 1786 and 1787. Alexander Mackenzie notes that a post (probably two posts) were established on Slave Lake to the east of the entrance of the Slave River by Leroux and Grant. It is probable that Leroux established a post under Ross and the opposition and Cuthbert Grant under Pond's direction, both erecting their buildings in the same locality. In the same year it is probable that a post was established on Peace river above Vermilion Falls referred to on Alexander Mackenzie's map as the old establishment. Pond's summer of 1786 was probably spent in organizing the district so that an ample supply of provisions could be obtained to enable the canoes to make the return journey to Grand Portage without difficulty and so that the trade and exploration could be developed to the largest possible extent. It was only through the establishment of an adequate base in the Athabaska district for provisions and furs that exploration could be carried on. During the winter of 1786-7 competition between Ross and Pond became more severe and news reached Roderic Mackenzie about the beginning of June that Mr. Ross "had been shot in a scuffle with Mr. Pond's men". The amalgamation of 1787 was arranged at Grand Portage between the small traders and the Northwest Company. Alexander Mackenzie received one share and was appointed to replace Pond who was apparently anxious to devote his final years exclusively to exploration. In the same summer Pond appears to have visited Slave Lake as he notes that in July 1787 there was a great deal of ice on that lake. It is not known whether he crossed over to the north side of the lake but he apparently learned that the Mackenzie river flowed from the western end and it is quite possible that he may have gone as far as the entrance of the Mackenzie. According to Ogden's report of a conversation with Pond the entrance of Mackenzie river was in Lat. 64.° Long. 135.° Alexander Mackenzie in 1789 was able to proceed directly to the entrance of the river.[11]

In 1787 Alexander Mackenzie arrived as his successor and Pond was obliged to leave Athabaska forever in 1788. During the winter however he apparently fired Alexander Mackenzie with the possibilities of discovery down the Mackenzie. There can be no doubt but that he gave Mackenzie all the necessary information and that Mackenzie's journey was the fulfilment of Pond's plans. Mackenzie was not at this time a successful trader or an explorer but he learned much from Peter Pond during the winter of 1787-8. He was given full instructions in the management of the new district. Alexander Mackenzie wrote in a letter dated Athabaska, Dec. 2, 1787 that McLeod and Boyer had been sent to the Beaver country on Peace River for provisions. Leroux had been ordered with Pond's concurrence to abandon Slave Lake. Masson suggests that in January 1788 Mackenzie had become convinced of the possibilities of the expedition to the Arctic. To his subordinate Roderic Mackenzie he wrote "I already mentioned to you some of my distant intentions. I beg you will not reveal them to any person as it might be prejudicial to me, though I may never have it in my power to put them in execution". The context however makes it difficult to decide that the "distant intentions" referred to the voyage down the Mackenzie river. From his superior Patrick Small, in a letter dated February 24, 1788, we learn of "the wild ideas Mr. Pond has of matters, which Mr. Mackenzie told me

---

[11] He was guided by one of the Indians but he was never in doubt as to his direction.

were incomprehensibly extravagant—He is preparing a fine map to lay before the Empress of Russia". Pond appears to have decided on the venture after a journey to Grand Portage in 1788. This may be inferred from a letter to Small dated December 3rd 1787. Small gave orders to him "to go with or after the packs, but represented to him that he required to be expeditious, if he intended returning after seeing the Grand Portage". Probably Pond decided to go down the Mackenzie in 1789 but arriving at Grand Portage in 1788 found it necessary to go to Montreal. It was at this point that he left instructions to "another man by the name of McKenzie—with orders to go down the river and from thence to Unalaska and so to Kamschatka and thence to England through Russia". Whether Pond left these instructions or not Alexander Mackenzie had reached the conclusion that Mr. Pond's ideas were not as wild as they had at first seemed and with Pond out of the country it was possible for Roderic Mackenzie to write in July 1788. "He (Alexander Mackenzie) then informed me, in confidence that he had determined on undertaking a voyage of discovery the ensuing spring by the water communications reported to lead from Slave Lake to the Northern Ocean". In a letter dated February 15th, 1789 he outlined to the partners of the Northwest Company plans for the voyage.

Pond came down to Montreal in 1788 and remained at Montreal and Quebec until March 1790 when he returned to Milford. He may have learned by the winter express by that time of the voyage of Alexander Mackenzie and realized that his conclusion formed during the past five years had been disproved. With Pond as with Alexander Mackenzie the Mackenzie River was the River Disappointment. Pond sold out his share in the Northwest Company in that year for £800 to Mr. William McGillivray and severed his connection with the Company forever. He was fifty years of age at this time and an old man for the fur trade. It was left for a young man to reap the rewards of his work and his experience.

V. THE MAN AND HIS WORK

Peter Pond was one of the sons of Martha. His achievements considering the handicaps[12] under which he laboured were in many ways remarkable but they were not of a sensational character. He was the first to cross the Methye Portage into, and to outline the general character of, the Mackenzie River drainage basin but aside from a little known map nothing has remained to show the importance of his work. He apparently wrote a journal but only after he had reached the age of sixty; and because of its inaccurate spelling it was only published a century later, after a large part of it had been destroyed, and chiefly as an example of orthography. He apparently had little training in astronomy and mathematics and worked with his instruments under great handicaps but he produced a map of very great value—the first map of the Canadian Northwest. It was one of the misfortunes of Peter Pond that the fur trade was productive of bitter enmities and that the chief chronicler of his activities in the Northwest should have taken the opportunity not only to neglect the importance of his work but actually to malign him. The hostility of the small traders to the Northwest Company was never overcome. Alexander Mackenzie was probably never a member of the Northwest Company with the full support and loyalty of its chief shareholders. His work was written after he had broken from them and his account of the fur trade was written from the standpoint of that small band of traders who never surrendered. History has

---

[12] Even David Thompson writes, "He was a person of industrious habits, a good common education, but of a violent temper and unprincipled character," *op. cit.* p. 172.

not been kind to Peter Pond. It has taken the word of the chief chronicler of his activities without question.

The intention of the foregoing sketch of Peter Pond's life is not to place him on a pedestal above the traders of his time. He had his faults. He was very proud and very sensitive. One needs only to read the remarks of his journal on Lahontan, and on Carver and of the making of a map to present to the Empress of Russia to realize his egoism but this egoism was partly the result of his age. He had achieved great things. It was unfortunate that he was unable to make people realize the extent of his achievements. He apparently died in poverty in 1807 without recognition when Sir Alexander Mackenzie was receiving the rewards of his own and of Pond's labours. His achievements have already been indicated. He was the first to penetrate to Athabaska and to suggest the lines of future exploration on the Mackenzie. He was the first to organize the trade as it was carried on to Athabaska. To him belongs the honour of having solved the problem of conducting trade over such long distances and of organizing the Athabaska department which was crucial to the development of the Northwest Company and to the prosecution of further exploration.

The task remains of indicating more specifically Pond's contributions to exploration. His map will probably remain an outstanding and permanent witness to his ability and energy. It is difficult to indicate the extent to which he had been trained in mathematics adequate to determining position but it is probable that a commission in the army may have required some training along these lines and that this may have been supplemented by information acquired on his two voyages to the West Indies. But an adequate training was not a substitute for accurate instruments. To gain an accurate knowledge of degrees of longitude it was necessary to have accurate chronometers and to maintain accuracy over long distances and for two or three years was extremely difficult. Pond had explored the St. Peters river beyond the limits previously reached by Carver or by any other white man but this exploration had no immediate importance. The penetration of the Athabaska district opened to explorers the possibilities of a new and large drainage basin. The results of Pond's first visit to the Athabaska country were given in Alexander Henry's letter dated Montreal, 18th October, 1781 to Joseph Banks. Pond probably had no instruments on his first expedition and the estimate of Alexander Henry based on Pond's information giving a location for Methye Portage of 60° N. latitude and 140° West longitude from London was very inaccurate. On the other hand Pond had acquired a fairly accurate conception of the character of the country. Henry suggested that a party should be sent down the river to explore the Arctic coast "which can be no great distance from the streight which seperates the two contenants." But he advised that the party should return "by the way of Hudsons Bay being much the shortest way back." This information based on Pond's work indicates that Pond believed at that date that the Mackenzie river drained to the Arctic Ocean. He returned to his old post in Athabaska in the summer of 1779, but came out to Grand Portage in 1780. He may have had instruments on this expedition but the results make this improbable. He was not able to return to Athabaska until 1783 and then was obliged to come out in the following year. He was compelled to acquire the necessary information on the district in these three intervals scattered over a period of four years. His chronometers,[13] if he were in possession of those instruments, would consequently be of slight value. According to Pond's map of 1785 Lake Athabaska is placed in longitude 130° whereas the Atlas of Canada for 1916 gives 112°. Lac La Ronge according to Pond is 105° longitude and according to the Atlas 116°. On the other hand his instruments for determining latitude did not suffer through the lapse of time and the Atlas of Canada gives for Lake Athabaska 58° and Pond's map 60°, and for Lac la Ronge 56° and Pond's map 55°. The problems of determining latitude and longitude are fundamentally different from the standpoint of obtaining accuracy over a long period of time. Although Pond's map was inaccurate it did show clearly the routes which were followed by the traders to Athabaska and the relative location of the important rivers, lakes and portages. It showed further the accuracy with which Pond was able to gather information from the Indians. "Il ne faut pas ecrire que les details que je mets sur cette carte ont aucune analogie précise avec l'endroit meme eloigne dans les bois avec peu de papier j'ecrivai mes reflexions et les placoir ou je pourais."

  The comments which accompanied the map and which have been printed in Davidson should be given more than a casual notice. In the first place he comments on the area south of 49° and east of the great plains the greater part of which is ideal country for beaver.

---

[13] David Thompson held that Pond used a compass "and for the distance adopted those of the Canadian canoe men in leagues." Thompson claimed that Pond reckoned the league as three miles whereas it should have been two miles and that this error occasioned most of the difficulty as to longitude. *David Thompson's Narrative, op. cit.* p. 72.

PETER POND: FUR TRADER & ADVENTURER

"Cette region peut veritablement etre appellie la region des castors." Secondly he estimates that the number of degrees from Churchill to Athabaska is 23° whereas his map shows approximately 35° and it is actually 17°. He concludes that it is 71° between Athabaska and Behring straits whereas it is actually 57°. "Je présume que la distance reelle ne peut pas être de plus de 60 dégrés." Above latitude 54° the country is thinly populated and the Indians live during the summer on reindeer, fish, and game. They trap the reindeer in nets and use the leather for clothing. In the winter they hunt beaver to trade with the English for rifles, powder and other necessaries. This applies particularly to the country above 58° and below that line the people depend on buffalo and moose. He divides the Indians of North America roughly into two groups east and west of a line drawn from 40° lat. 95° long. to 60° lat. 13° long. or according to his map a line roughly following the edge of the Canadian Shield from the Mississippi to Lake Athabaska. All the tribes east of this line, (the Algonquin family) speak a language similar to that of the Eskimo in the Labrador coast, are very adept in the construction and navigation of canoes and are friends and allies among themselves as far as 60° North. The tribes west of the line speak "un langage extraordinaire, que consiste dans un bruit de gosier qu'il est impossible d'aprendre." They are ignorant of the canoe, are allies and friends among themselves but constantly at war with the Indians of the first group. On the death of the parents the western Indians cut off one of their fingers while those of the east pinch the skin of the arms and legs and pierce them with the point of a knife to make them bleed. The tribes of the east are steadily gaining on those of the west although the latter attempt to break through but they are forced back because of the possession of fire-arms by those in the east. The contrast between the two peoples forced him to conclude that the Eastern part of the continent was settled from Europe and the Western from some part of Siberia. The constant warfare may have been due to these discrepancies. In any case the people of the west had been forced back a great distance during the past 40 years.

These comments appear to describe very accurately Pond's reflections on the Indians and on the country. They are taken from a memoir supposedly written by Pond which also contains much other information. With this memoir is included an extract covering much the same material but according to Davidson written in a different hand and using much better French. The second memoir has also rearranged much of the material confusing it and in some places making it clearer. From the two accounts it becomes apparent that Pond was interested in gaining information from all the natives he had met. The words "a Arabosca le 15 mars 1784" are given a different place in the memoirs. In the second memoir they refer definitely to the visit of the Copper Indians on that date. These "gens du couteau rouge" lived a short distance from the North sea and had a large number of copper knives. Pond purchased some of these and brought them down with him. They told him the sea "etoit sujette au flux et reflux que ses côtes couroient fort loin vers l'ouest, et qu'ils ne connoissent aucune terre au nord: que les rivieres de leur pays couroient au Nord ouest et alloient de jetter dans cette mer; que toutes les rivieres etoient navigables des le commencement de l'été et qu'ils avoient vu plusieurs fois des isles de glace flotter sur leurs cotes: mais ce qu'ils ajoutoient qu'il n'y a point pendant l'été d'obscurite totale dans leur pays, mais toujours une espece de Crepuscule, me fit croire que cette nouvelle mer est sous le cercle Polaire Arctique." They apparently could not see the midnight sun and Pond concludes they were under the Arctic Circle. The meeting with Hearne's Indians is again described without change. The country north of Slave Lake he describes as an immense plain "dans lesquelles il n'y a ni bois, ni herbe; on y trouve seulement quelques buissons dont le grosseur n'excede pas celle de la jambe. Tous les sauvages du Nord sont gens fidels et honnetes." It was through this country that Pond expected they would find the Northwest Passage. Under the 60th degree of latitude and around the 125th degree of longitude or south of Lake Athabaska the rivers are full of ice, from the beginning of May whereas those which run into Hudson Bay are not open until the middle of June and this in spite of the more southerly location of York Fort and Churchill Fort. Under 60° "on trouve tous les fruits naturels a l'Amerique et au climat tels que les fraises, groseilles etc et une infinite d'autres pour lesquels je n'ai pas de noms. Ils croissent sur les bords des rivieres et des lacs qui abondent en poisson, et sont couverts d'une multitude infinie d'oyes et de Canards sauvages. Dans ce canton quand le vent est a l'ouest l'atmosphere est convert de nouages et de brouillards humides meme pendant l'hiver, mais au contraire quand le vent est sud-est tout est clair et serein." Pond later proved that the climate was adapted to agriculture by his garden. Alexander Mackenzie wrote that when he arrived at Athabaska in the fall of 1787 "Mr. Pond was settled on the banks of the Elk river where he remained for three years and had formed as fine a kitchen garden as I ever saw in Canada.—In the summer of 1788 a small spot was cleared at the old Establishment, which is situated on a bank thirty feet above the level of the river, and was sown with turnips, carrots and parsnips. The first grew to a large size and the others thrived very well. An experiment was also made with potatoes and cabbages the former of which were successful; but for want of care the latter failed."

These memoirs also include information acquired from Indians and elsewhere regarding the character of the country west of the Rocky Mountains. Pond suggests that he had seen the mountains "J'ai aussi vu dans ces montagnes des pierres a fusil, pleines de veines d'un metal blancs; je ne sais si c'est de l'argent, ou ce que ce peut-etre." In this uncertain use of the French language this should probably be interpreted as meaning that he had seen gun flints brought from these mountains full of veins with a white metal. There is no other indication of his having been as far west as the Rocky Mountains. He had met Frenchmen who had lived among the western Indians as in the case of old Pinnashon at the Wisconsin Portage, a Frenchman who "impose upon Carver respecting the Indans haveing a rattel snake at his call which the Indans could order into a box for that purpos as a feat. This frenchman was a solder in the troops that ware stasioned at the Elenoas. He was a Sentanel. At the Magasean of powder he deserted his post and toock his boate up the Miseura among the Indans and spent many years among them. He larnt maney langwedgeis and from steap to steap he got among the Mandans whare he found sum french traders who belonged to the french factorey at fort Lorain on the Reed River. This factorey belonged to the french traders of Cannaday. These people toock Pinneshon to the factorey with them and the consarn toock him into thare sarvis til the hole cuntrey was given up to the English and he then came into thare sarvis. The french strove to take him up for his desarson but fald. However they ordered him to be hung in efagea which was dun. This is the acount he gives of himself. I have heard it from his one lips as he has bin relateing his adventures to others." Pond's attitude toward "Pinneshon" explains the accuracy of the information which he obtained from Indians and others. He was a shrewd judge of human nature. He complains that this man "found Carver on this spot going without understanding either french or Indan and full of enquirey threw his man who sarved him as an interptar and thought it a proper opertunity to ad sumthing more to his adventers and make his bost of it after which I have haird meney times it hurt Carver much hearing such things and putting confadens in them while he is guvner." His appraisal of Carver is worth noting. "He gave a good a count of the small part of the western country he saw but when he a Leudes to hearsase he flies from facts in two maney instances," Pond took no stock in stories of pet rattlesnakes.

His account of the country west of the Rocky Mountains could not be accurate but it illustrates again Pond's ability to secure information. These mountains were called by the Indians the Rocky Mountains and "en plusieurs endroits elle est taillée a pic". They extend the full length of the plains and at places are as wide as 150 leagues. Along the mountains there were "sources d'eau bouillante, d'autres d'une chaleur plus modérée et un nombre infini de froides." The Indians apparently explained in detail the character of Peace River. "Les naturels disent encore qu'il un chasme ou passage souterrain à travers cette montagne par ou coule la riviere qui vient du Lac Arabosca, que les rochers au dessus sont perpendiculaires et d'une hauteur immense et qu'ils ont osé pénétrer à une petite distance dans leur canots, sous ces voutes terribles." The description probably applies to the Peace River below the junction of the Finlay and the Parsnip. "Ce qu'il y a certain c'est que le long des bords de cette riviere et du Lac Arabosca on trouve des sources de bitume qui coulent sur la terre." There could be no doubt as to the reports of coal and tar.

On the other side of the mountains "toutes les rivieres courent vers l'ouest et vont se jetter dans la mer du sud suivant ce que m'ont dit les sauvages." On that side there are many plains which leaving 45 degrees of latitude "et allant vers de midi, sont fort chaudes dans l'été." The tribes of these plains have nothing in common with those on the other side of the mountains "avec lesquelles sont toujours en guerre. Toutes ces tributs de l'ouest sont tres unies entr'elles et en general tres simples et de moeurs fort douces." These tribes did not use canoes and having no knowledge of fire-arms were very much at a disadvantage in war with the Indians east of them. Their language differs from those on the plains although they cut off a finger on the death of the parents. These tribes extend to the Pacific Ocean where the eastern tribes pursue them and take them prisoners. "J'ai meme vu plusieurs chevelures de negres qui avoient été tués dans ces rencontres" a statement for which one can find no explanation. The animals are different from those of the plains and resemble those of "l'Amerique meridionale" especially the lama with its beautiful wool. The Indians situated near the 48th degrees raise "un excellent tabac dont le saveur est toute particuliere." They have also horses, some mules, and asses, and great troops of buffalo from 54° to 63°.

PETER POND: FUR TRADER & ADVENTURER

Pond had gained as shown in these comments a surprising knowledge of the general topography of the North American continent. He had during his experience as a trader visited the Mississippi, Lake Dauphin, the North Saskatchewan and the Athabaska and from these rivers of three drainage basins he had obtained a thorough grasp of the character of the country and of its inhabitants. On his visit to Montreal in 1785 he had come in contact with the volumes of Cook's voyages, possibly W. Ellis, *An authentic narrative of a voyage performed by Captain Cook and Captain Clarke*, (London, 1782). Cook's inlet is shown on his map and also Prince William sound and King George's sound. The references in his memoir to Cook's voyages are few and were probably added to the notes he had made in the interior. He notes that the Indians west of the Rocky Mountains cut off a finger on the death of their parents similar to those inhabitants of Middlebourg in the Friendly Islands of the Southern Pacific mentioned by Capt. Cook. The location of Behring Strait, of Cook's inlet and other points on his map were borrowed directly from the volumes on Cook's voyages. The map of 1785 is the most enduring testimony of Pond's contribution to the geography of North America.

He returned to Athabaska with a detailed knowledge of Cook's voyages and with plans to bridge the gap of territory to the Northwest which was still unknown and also to verify the information he had obtained from the Indians. Prior to his departure the Montreal merchants had become thoroughly aroused as to the possibilities of the fur trade on the Pacific coast and especially in the territory served by Cook's river. In Pond's memorial to the Hon. Henry Hamilton dated Quebec, April 18, 1785 he writes that "he has had positive information from the natives who have been on the coast of the North Pacific ocean, that there is a trading post already established by the Russians; and your memorialist is credibly informed that ships are now fitting out from the United States of America, under the command of experienced seamen (who accompanied Captain Cook on his last voyage) in order to establish a fur trade upon the Northwest coast of North America, at or near to Prince Williams Sound and if the late treaty of peace is adhered to respecting the cession of the upper posts, the United States will also have an easy access into the Northwest by way of Grand Portage. From these circumstances your memorialist is humbly of opinion that this branch of trade will very soon fall a prey to the enterprizes of other nations, to the great prejudice of His Majesty's subjects, unless some means are speedily used to prevent it. It therefore becomes necessary for government to protect and encourage the Northwest Company in the earliest prosecution of the proposed plan; in order that trading posts may be settled and connections formed with the natives all over that country even to the sea coast; by which means so firm a footing may be established as will preserve that valuable trade from falling into the hands of the other powers; and under proper management it may certainly in a short time be so extended as to become an object of great importance to the British nation and highly advantageous to this mutilated Province." Alexander Henry, who had probably been in frequent consultation with Pond wrote to his friend William Edgar, who at that time lived in New York, having made substantial profits in the forwarding trade from Montreal to Detroit during the American Revolution, on Sept. 1st., 1785 regarding the trade on the Northwest Coast. "By yours (of the 15th, July) I find you intend to become an adventurer in Chinna trade. I think it will answer until overdone, which will soon be the case were the last adventurers are fortunate, my scheme for the North coast of America I think will soon take place as I am told they are fitting Albany sloops for Chinna, they may as well send them to Cook's river where I am persuaded they will receive more profits than from all the upper posts." In a later letter of March 5, 1786, Henry makes his plans more specific. "Montreal, 5th March 1786. Dear Edgar, I find you are largely adventuring in the East India business, I am not the least doubtful of its proving beneficial I also observe you have some idea of putting my favourite plan into execution, of carrying the trade to the North west coast of America, it is a shame for America to let slip such a valuable trade and extensive territory which nature has given them the best title to, the world in general sees it at present what I acquainted you of two years ago, and I doubt not but the Empress of Russia will make settlements from Kamschatka soon, there is a long extent of country from Cape Blanco to Cook's River, extensive Bays, and large rivers empty or fall in the sea, establishments should be made at the enterance of each as far up as shipping could go, and a small fort built at each for the protection of those remaining which could be done at a very moderate expense, as we are obliged always in the North to build Stockade forts, such establishments if proper persons conduct the business in a short time would bring numerous tribes of Inland Indians with the finest furs in America and in great quantities exclusive of the sea beaver and sea otter skins sells at such great prices in China, whose numbers would be increased by their having other means of destroying them than what they have at present, as for provision they would be easily procured from the abundance of fish and wild fowl with which all Northern countrys abound but more particularly the rivers Cook entered, I make no doubt but Cook's River (called also Sandwich Sound) has a communication with those parts of the Northwest I was at, by which a road would be opened across the Continent and in the end might produce great discoverys which would prove beneficial to the world and society in general were I without a family which must stay with and provide for I would set off immediately for there is no one I could recommend so well as myself, nor no voyage could please me so well, but should you be serious in your intentions (a better scheme, I am sure you could not undertake) let me know and I will perhaps find some person which may be of service to you. In the postscript of a letter Oct. 22, 1787, Henry writes "I am informed there is an expedition to the South Seas from Boston. I hope you will have the honour of being concerned as it will redound much to those who are the first undertakers, let me know particularly about this matter."

The discovery of a large river by Cook emptying into the Pacific (Cook's Inlet) was the occasion for much speculation. Alexander Henry was convinced that the Athabaska river eventually found its way to Cook's Inlet. Pond also became convinced that the Mackenzie river emptied into the Northern Pacific. Whether he arrived at this conclusion before he left for the Northwest as a result of conversations with Henry and others or whether he arrived at it during the later years in Athabaska cannot be determined. It represented a reversal from the opinion set down in the map of 1785.

On his arrival in 1788 he was the object of much interest. There can be no doubt that he had persuaded himself, possibly after his visit to the entrance of the Mackenzie in 1787, and Alexander Mackenzie that the river flowing from Slave Lake was that which emptied into Cook's Inlet. Mackenzie wrote to Lord Dorchester on November 17th, 1794 that he had followed "the course of the waters which had been reported by Mr. Pond to fall into Cook's river, they led me to the Northern ocean." A part of Mackenzie's journal cited in Davidson from the Stowe MS notes "It was in the summer of 1789 that I went this expedition in hopes of getting into Cook's River, tho' I was disappointed in this it proved without a doubt that there is not a North west passage below this latitude." In a letter to Roderic Mackenzie at Fort Chipewyan, March 2, 1791, he refers to the "river Disappointment." In a tirade against the English chief on his return journey he writes "I stated to him that I had come a great way, and at a very considerable expense, without having obtained the object of my wishes." But he did learn of the Yukon river on the other side of the mountains and from constant inquiry of the Indians concluded that the fort which had been described as at the mouth of the river was "Unalaska Fort and consequently the river to the west to be Cook's River. I made an advantageous proposition to this man to accompany me across the mountains to the other river but he refused it." He realized that the voyage was a failure and in a letter to Roderic Mackenzie dated Grand Portage 16th July 1790 he wrote "my expedition was hardly spoken of but that is what I expected."

    Pond's actual report on the river has not been found and information regarding his views must be obtained at second hand from a letter written by Isaac Ogden at Quebec dated 7th November, 1789 to David Ogden in London describing conversations on the subject. The report is extremely difficult to follow but its comments are suggestive as to Ogden's confusion. He describes Grand Portage which leads to the waters of the Northwest, and the Mississippi which is reached by another portage from the head of Lake Superior, and the navigation down the Mississippi to the mouth with no interruption except St. Anthony's Falls. It is difficult to understand what is meant by the sentence "the traders go on this course westward leaving the Mississippi to the eastward one thousand miles" but presumably he means west one thousand miles from Grand Portage. "The furs in this district are much inferior to those from the Northwest posts." From the end of the portage at the head of Lake Superior all the lakes and waters as high up as Lat. 58° and long. 124° (probably Methy Portage) set first to the Northwest and North and then take a south easterly and south course and empty into York (Hudson's Bay) probably meaning that the Red River flows to the north and the Saskatchewan to the south although this is difficult to imagine. The Canadian traders pass one of the Hudson's Bay Company's post at lat. 57° long. 110° (Cumberland House?) and proceed to Methye Portage and Athabaska in which the rivers drain to the north into Slave Lake and the Northern ocean. The lakes emptying into Slave Lake include the Arabaska, the Lake of the Hills (which is the same) and Lake Pelican which is difficult to identify. The Slave river carries the waters of these lakes to Slave Lake. It runs northwest and is several hundred miles long. A very large river leaves Slave Lake at 64° lat. and 135° long, (out nearly 20° on long. and probably 3° on lat.) and it has one of the largest falls in the world at 141° long. It is possible that Pond may have visited the entrance of the

Mackenzie river and taken observations and learned from the Indians of the falls on the Hay River which empties into Slave Lake near this point. The Mackenzie River leaves the lake in a south westerly direction. Ogden pointed out that the Rocky Mountains terminated in 62-1/2° lat. and 136° long. and that the Slave River running to the westward of them emptied into the Pacific in lat. 59° or at the mouth of Cook's river. From Cook's voyages he concluded that the river emptied in 59° 40 lat. and 154° long. W. Since Cook had explored the river for 70 leagues it was concluded that there remained a very short stretch to be explored. The amount of drift wood found in Cook's river could only have accumulated on the Slave river. Two Indians had brought a blanket to Slave Lake from the mouth of the river in 1787 and there could be no doubt that the Northwest Passage had been found. On July 15, 1787 there was a great quantity of ice on Slave Lake and the Indians penetrated to the Arctic in that year and killed some of the Eskimos. Cook had gone as far as 72° and was stopped by ice therefore Ogden believed and rightly that the coast extended to the south from this northern point to 68-1/2°. Ogden summarizes his conclusions and reaches the significant results "that an easy communication with, and an advantageous commerce may be carried on by posts established on Lakes Slave, Arabaska, Pelican, etc., etc. and to deliver the fruits of their commerce at the mouth of Cook's river to be then carried to China, etc. and that as Cook's river and the lands on Slave Lake, Arabaska, etc. are very fine, some advantageous settlements may be made there which may be beneficial to government." Mackenzie had been left by Pond to go down the river to make a report.

These were Ogden's observations on Pond's conversations. In the main they were probably accurate although Pond appears to have reported to Captain Holland that the Slave river took its course from Slave Lake to the Northwest and to Prince Williams Sound. On his arrival at Montreal and Quebec in 1788, Pond and other members of the Northwest Company appear to have thrown their energies into the project of securing government aid for a discovery of the river. To the traders of the Northwest Company at Montreal, Pond's information raised hopes of a Northwest Passage, of access to the profitable Pacific trade and of a short route to the important Chinese market for the furs of the Northwest—all of these under their control. The competition on the Pacific coast which followed the publication of Cook's voyages from the Russian and the American and other foreign vessels was an additional incentive to government exploration. Captain George Dixon who had just returned from a very profitable voyage on the Pacific coast wrote to Evan Nepean on July 14, 1789; "if something is not done and that immediately this valuable branch of commerce will be lost to this country and in consequence of that loss the traders both from Hudson's Bay and Canada will find themselves in a bad neighborhood."

The Hudson's Bay Company was also alert to the possibilities of this new route falling into the hands of the Canadian traders. Ogden's letter to his father was forwarded to Evan Nepean on Jan. 23, 1790. Alexander Dalyrymple, later (1795) the hydrographer of the Admiralty, presented a memorandum on Feb. 2nd, 1790, discrediting Pond's observations but preferring the despatch of two vessels one "round Cape Horn without delay and another to Hudson's Bay and the Hudson's Bay Company have expressed their readiness to cooperate with government." Dalyrymple reported on Feb. 11th, 1790 that "My Friend Mr. Wegg, the governor of the Hudson's Bay Company, desires me to say that the Directors of that Company have unanimously determined to send their sloop of about 90 tons at the Company's expense if government will send a proper person in her to examine if any outlet can be found from Hudson's Bay to facilitate the communication with the West Coast. They are particularly solicitous that government would send a proper person in her that the publick may be assured of everything being done to effect the desired purpose. They also wish that two proper persons may be sent by government to travel inland to ascertain the shortest communication by the lakes and rivers and the Hudson's Bay Company will defray any reasonable expense of that undertaking." Indeed the Company despatched Philip Turner[14] to Athabaska in 1791 to determine the locations on the new route. The Company had already received Hearne's report and probably expected little from the journey but they were anxious to get all the information possible. Nevertheless the Canadians were not to be thwarted and Captain Holland worked out a plan with considerable detail for the exploration of the land lying between "Lake Aurabusquie—and the line of coast discovered by Cook." As late as July 25th, 1790 Captain Holland in a letter to Evan Nepean suggested plans for the prosecution of exploration in the following year. The news of Alexander Mackenzie's journey to the Arctic brought these plans to a sudden end. A map of T. Conder published in London on Jan. 1st,

---

[14] See *David Thompson's Narrative, op. cit.*, pp. 173-4.

PETER POND: FUR TRADER & ADVENTURER

1794 included for the first time the route followed by Alexander Mackenzie down the Mackenzie to the Arctic and gave the first outline of the Arctic coast from the mouth of the Coppermine to Icy Cape.

Pond had retired to Milford. On Monday, March 8, 1790 he made a first visit to Ezra Stiles the President of Yale University and from items in Stiles' diary additional information may be gained as to Pond's views although again these are by no means clear. He estimated that his house and settlement (presumably the old establishment on the Athabaska River) was located thirty days west of Hudson's Bay and in the 60th degree North latitude. He claimed to have been "within six days travel of the Grand Pacific Ocean or the western side of N. America", or possibly the entrance of the Mackenzie River. On March 24th the diary notes that Pond had resided three years—in 64th deg. of N. Lat. (at the right of the entrance of Slave River in Slave Lake) information verified on Ezra Stiles' copy of Pond's map. He described the various traits of the Indians stressing the general similarity of Indian culture throughout northern North America. According to his account there were over twenty trading posts beyond Lake Superior. He expressed the opinion that Lord Dorchester, the Governor of Quebec, was anxious to conceal all the discoveries in the Northwest and to monopolize the fur trade. Finally Pond contributed geological specimens to the museum of Yale College. He apparently left a copy of a "large map of his own construction" since the copy found among the Stiles papers is dated March 25th. But Stiles was engaged in copying the map at least until April 7th. On Sept. 15 President Stiles paid a return visit to Pond at Milford. The map copied by Stiles as already suggested provides an excellent basis for a study of the posts and the trade routes but has serious limitations as to latitude and longitude. On the left hand corner Nootka is located at 130° longitude and 49° latitude but Pond does not include a map of the prospective outlet of Slave Lake through a river to the Pacific Ocean. After learning of Mackenzie's failure he probably destroyed the large map which he had shown to Ogden and submitted a rough revised map to Stiles. His reversal of opinion after the map of 1785 which was the occasion for numerous projects to prove the existence of a direct line of communication from Athabaska river to Cook's inlet and which Alexander Mackenzie disproved had a sad ending. He left Canada with all his later conclusions disproved. Had his conjecture proved correct he would have been accorded a place among the great discoverers of Canadian history. But it was proved wrong and his former friends and supporters probably regarded him as a traitor. Alexander Mackenzie found that he was also mistaken but youth was on his side and he lived to make the journey to the Pacific by the Peace River. The latter achievement offset the disappointment incidental to the voyage down the Mackenzie river.

He apparently spent the remainder of his days reading the works available on the Northwest including Lahontan and Carver. Later accounts attribute to Pond considerable bitterness which resulted from the lack of recognition for the great services which he had rendered. There are suggestions that he was disappointed with the Northwest Company and that he gave information to the United States government in the settlement of boundary disputes but there appears little to substantiate these charges. One would not have been surprised to find this bitterness but his journal which was apparently written after 1800 is the most direct evidence to the contrary. Peter Pond had a full life and he had much to think of in the last days of poverty. One cannot forget the old man noting down his story, becoming confused as to the sequence of events, with so much to write and such difficulty in writing it, remembering the campaigns of the Conquest, the ginger bread and "small bear", the cold tents, the fighting at Ticonderoga under Abercrombie, his old comrades in arms, his trips to the West Indies, the duel, the trade, on the Mississippi and in the Northwest, chuckling to himself as he remembered the stories of the trade, the Indians, and the voyageurs, remembering with the detail which always comes with physical effort, the good and fat ducks, the warm ground, his garden at Athabaska, the pull upstream of the canoes, the portages, the gentle gliding stream "as we desended it we saw maney rattel snakes swimming across it and kild them."

Canadians must ever remain grateful to Pond for his pioneer work in organizing the fur trade to Athabaska, and for his active part in the development of the Northwest Company which was the precursor of the present Confederation. Above all, the most experienced fur trader and in that an explorer and an organizer, he laid the basis for the later exploration for Northwestern Canada. Like the other traders from the colonies Alexander Henry, Simon McTavish and the Ellices he felt no strong allegiance to any government but allegiance to Great Britain was a prerequisite to a supply of manufactured goods essential to the fur trade. Capital, skill, and connections with English supply houses were brought from Albany to Montreal to eventually establish the Northwest Company which became one of the important influences in the establishment of Canadian unity from the Atlantic to the Pacific and in the maintenance of Canadian allegiance to Great Britain. Peter Pond did important work in organizing the Athabaska department which became the raison d'etre of the Northwest Company and the keystone by which it was able to extend its operations from the Atlantic to the Pacific. If lessons were to be drawn from his life, nothing would be more obvious than the fruitlessness of sentimental lamentations over the weakening of the ties of the British Empire. The Empire has grown and been maintained on stronger bonds than political bonds and it has grown in spite of its builders as well as because of them.

BIBLIOGRAPHY

The works dealing with the activities of Peter Pond are extremely scanty and in most cases very unreliable. The only record from Pond's own hand is a manuscript of a journal, a large part of which has been destroyed and the remainder of which is in the hands of Mrs. Nathan Gillet Pond. This remnant was printed in the *Connecticut Magazine*, Vol. X pp. 239-259 and reprinted in the *Wisconsin Historical Collections*, Vol. XVIII, pp. 314-354. A reprint of a small part of the journal was given in the *Journal of American History*, Vol. I, pp. 358-365. The manuscript was written probably after 1800 but the details which have been checked by the editor, the late R. G. Thwaites leave no doubt as to its accuracy and authenticity. I have included a large portion of this journal, rearranging it to give proper sequence and omitting less important details, because of its very great importance to a study of Peter Pond as well as of the fur trade. The extent to which I have been indebted to the Wisconsin Historical Collections will be evident to any student in the field. A memoir written in French and accompanying a map presented to Congress in 1785 published in G. C. Davidson, *The Northwest Company* (Berkeley, 1918) pp. 259-266, affords some light on his later activities but otherwise we are dependent for information on data supplied by his contemporaries. A memorial dated Quebec, 18th April 1785 was presented to the Honourable Henry Hamilton by Peter Pond on behalf of the Northwest Company but the voice is the voice and indeed the hand is the hand of B. and J. Frobisher. The information which he gave to contemporaries was most unsatisfactory and it is not difficult to imagine the old trader recounting his experiences, becoming confused as to dates and events and leaving his hearers in a veritable maze.

PETER POND: FUR TRADER & ADVENTURER

The first of these accounts is given in a letter from Alexander Henry to Joseph Banks, Montreal, 18 October, 1781 printed in L. J. Burpee, *Search for the Western Sea* (Toronto, 1908) appendix. Sir Alexander Mackenzie who wintered with him in Athabaska was unable to straighten out the sequence of events as is shown in his account *A general history of the Fur trade from Canada to the Northwest* printed with the voyages, (London, 1801). Isaac Ogden "after several conversations with the map before me" wrote a description of the northern interior as given by Pond, in a letter dated Quebec, 7th November 1789, to David Ogden, London which was published in the *Report on the Canadian Archives* 1889, pp. 29-32, but the account leaves much to be desired.

His reports to President Stiles are given in a sketchy manner in the *Literary Diary of Ezra Stiles, D.D., LL.D. President of Yale College*, ed. by F. B. Dexter (New York, 1901) Vol. III, pp. 383, 385-6, 388, 402. A map was presented to the Hon. Henry Hamilton, Lieutenant-Governor of the Province of Quebec dated 18th April 1785 and printed in the *Report on the Canadian Archives*, 1890, p. 52. Mr. L. J. Burpee has kindly shown me a letter dated Jan. 8, 1907 from P. Lee Phillips of the Library of Congress, Washington, D.C., expressing the opinion that this map or "the lithographed reproduction" is a copy of the original and that the Kohl copy preserved in the Archives of the Hudson's Bay Company in London and printed in L. J. Burpee, *Search for the Western Sea* (Toronto, 1908) p. 102 and in G. C. Davidson, *The Northwest Company* (Berkeley, 1918) p. 32 is a copy of a map by Crèvecoeur which in turn was a copy, with various changes and omissions, of the original. A report of the Coast and Geodetic Survey Office, Jan. 16, 1907, disputes this however and suggests that the lithograph map is a later edition of the Kohl map. Davidson located two copies in the British Museum, one of which he printed (p. 37). There appears little doubt but that the map which he was supposed to have been preparing in Athabaska to show to the Empress of Russia was the map which he showed to Ogden. With the news of Mackenzie's failure this map was doubtless destroyed and a revised map was made. A copy of the revised map was made by Stiles, President of Yale University, March 25, 1790, and is now in the possession of Yale University library. It again illustrates the confusion in which interviewers of Peter Pond were left. Notes were written on the map which obviously conveyed wrong information. Pond is said to have prepared another map of the Northwest at this period but no copies appear to be extant. An interesting controversy as to the date of the 1785 map will be found in L. J. Burpee, *Search for the Western Sea* (Toronto, 1908) ch. VII. and G. C. Davidson, *The Northwest Company* (Berkeley, 1918) pp. 36 ff. and especially the notes on pp. 37-8.

There are numerous later works which deal with Pond but most of them are unsatisfactory and based on the material cited generally without a critical appreciation. It would be possible to go through account after account and point out the errors which have persisted. One cannot afford to depart from the primary material and even this must be used with great caution in a study of the life of Peter Pond.[15] *The report of the Canadian Archives* 1889, pp. 29, ff. has reprinted some valuable documents on later history. *David Thompson's Narrative* ed. J. B. Tyrrell (Toronto, The Champlain Society, 1916) may be cited as a work which should be used with caution in the reference to Pond. Thompson joined the Northwest Company some time after Pond had left it and he can scarcely be regarded as a primary source. Stiles *Itinerary* vol. 6, 1791, pp. 49, 406-7 gives additional information as shown in the following extracts.

"Rev. Mr. Prudden of Enfd. was born at Milfd. & well acquainted with the Pond Famy. there. The Boys were all enterprizing, bold & adventurous. When the last Fr. War begun wc. was 1754 & 1755, they would list one after another in the Army. Peter Pond might be then 17 or 18 AEt. & so born about 1737 or 1736. He rose to an Ensigncy by the Peace 1763. In the war he & his Brothers had become acquainted with the Wildness & Indian & Fur Trade. And after Peace 3 or 4 Brs. were concerned in Fur trade among Indians. Peter went a Voyage to W. Ind. About 1766 he went into the Ind. Country & tradg in furrs was absent from his Wife & family seven years. In connexion with Brs. a Trade thus:—They sent down Furrs in Hudson R. to N. York—sold them for goods in Fall—came & traded out the Goods at Milfd. in the Winter & made remitta. to N.Y. In Spring took up Goods suitable for Ind. Trade & spent the Summer in Ind. Countries. Thus circuitously till Br. Zecha. was cut off by Indians. After Absence of y. Peter visited his Famy. at Milfd.—then 1773 went off again to the Sources of Mississippi & spent there & on the N.W. Waters to 64th.

---

[15] Roderic Mackenzie wrote in his notebook "Peter Pond, became rather ancient in the N.W. He retired in 1788. He thought himself a philosopher and was odd in his manners. I understood he published something of the Northwest. He died poor."

deg. Lat. seventeen years—employed in a System of a Fur Trade Compa. at Quebec which kept up 21 Tradg. Houses guarded by 800 West of L. Superior almost over to Western Ocean. In March 1790 Capt. Peter Pond returned again to his country & Famy, at Milford." pp. 406-7

"Major Sheshan acquainted with Sir Peter Pond in Montreal 1784 or 1785.—Knew him a sailor in Mr. Morgans Vessel at Killingwth—Saw Capt. Charles Pond at N.Y. 1785 or after—Charles told him his brother Peter killed Gen. Arroldy. Mate quarreled with him, kept another rifle, fought another duel with the Captain—fled to Canada by a Vessel bound from W. Ind. up St. Lawrence—that the Family at Milfd afterwards recd a Lett. fr. Canada informing that Peter was killed by the Indians." (p. 49.)

A further reference is found in *Saint John de Crèvecoêur sa vie et ses ouvrages* (Paris, 1883) pp. 108-9. "Crèvecoeur fit aussi, en 1785, parvenir en France la carte manuscrite d'un voyage dans l'intérieur du continent jusqu'à Arabosca (63° de latitude)".

Since this volume went to press the Detroit Public library has discovered and drawn to my attention various accounts and memoranda in the Williams Papers relating to Peter Pond. Mr. R. H. Fleming has also copied for me one or two documents in the Phyn Ellice papers in the Buffalo Historical Society library. In a letter dated November 17, 1771, in the latter papers, we find the statement "enclosed you have also Graham and Pond's draft on Felix Graham from Phyn Ellice and Porteous for £50.15,—Graham and Williams, draft on Felix Graham £6.6.8. Graham and Pond's note on demand £38. Graham and Pond for freight of 235 gallons rum and 25 bu. corn." It is certain from this that Pond was in partnership with Graham in 1771 and that the partnership of 1773 was simply a renewal (p. 20). The goods belonging to this partnership are described in detail in "An invoice of sundries received from Felix Graham on account of the partnership of Graham and Pond, Michilimackinac 30th. June, 1773," valued at £1244.13.11½ (p. 22). A document headed "Goods Left with Mr. Boban" (Beaubien) in Pond's handwriting and apparently dated December 20th, 1773, includes three bales of goods. Beaubien may have been his French competitor but was more probably a clerk. The returns for this venture (1773-4) are indicated in an "Invoice of peltreys delivered to Felix Graham by Peter Pond to be sold on ye act of Graham and Pond marked and numbered as pr margen Michlemekenac

3 July, 1774," including 31 packs (p. 45). The partnership with Graham was brought to an end, and an account beginning July 28th, 1774, is headed "Leidger Pond and Williams 1774-5" (p. 47). The new firm sold a small quantity of furs to Graham in 1774. A partial list of the goods bales Nos. 11 and 16 and additional pieces probably belonging to the outfit for 1774-5 and valued at £778.14½ is included among the documents (p. 47). Another document dated Le Bay, 22 May, 1775, "Memorandum of goods received from Mr. Shaney" (Chenier) valued at £1008.12.6 and signed by P. Pond provides an indication of the extent of the trade in the Mississippi. The furs acquired in the venture of 1774-5 are shown in the ledger under date June 22, 1775 (p. 67). The "Leidger Pond and Williams" 1774-5 shows Peter Pond Dr. June 22 and July 3rd. for the outfit probably taken to the Northwest (p. 69). It includes a beaver hat bought on June 14, 1775, and a small red trunk. Alexander Henry writes that he left the Sault June 10th, but according to the Askin papers 202 lbs. of grease were delivered to him June 15th (p. 69). In the Porteous papers in the Buffalo Historical Society library an account is headed "Adventure to the N.W. per Messrs. Pond and Greves" dated Grand Portage 22nd July, 1775, and includes 6 new canoes—probably 2 large ones brought to the Portage and 4 small ones taken to the North. The firm Pond and Greves was apparently a subsidiary of Pond and Williams. A document entitled "An account of goods debts and notes in the hands of Thomas Williams belonging to Pond and Williams dated Michilimackinac, 16th. July, 1775," gives an inventory of the latter's possessions (p. 69). The partnership of Pond and Williams dissolved in 1777 and was replaced by that of Pond and McBeath, apparently on April 17, 1777 (p. 36). These documents serve to render more certain and accurate the discussion of the events in the period 1775 to 1778. Other documents show that Pond probably returned to Athabaska in 1779 for the furs left in cache and that he came out in 1780 (H. A. Innis, "Peter Pond in 1780," *Canadian Historical Review*, Dec., 1928). The firm of McBeath, Pond, and Greves was granted a license to Grand Portage in 1781. The discussion following p. 82 should probably be corrected accordingly. Photostat copies of the above documents may be consulted in the University of Toronto library.

Regarding the murders of Waden and Ross of which Pond has been accused, the evidence against him is extremely slight. In "A report of the special privy council to consider its powers to try cases of murder in the Indian territory and a number of cases so tried including that of Francois Nadeau and Eustache Le Comte for the murder of John Ross at Arabaska" Q. 36, -1, 276-310, there is no mention of Pond. The report dated June 9, 1788 includes minutes of proceedings from May 20th, 1788 to May 29th, 1788 and represents a first inquiry into the whole question of the right to try cases of murder in the Indian territory. There is no mention of the murder trial of Waden as establishing a precedent and while this is not conclusive evidence of Pond's innocence it does show that evidence of his guilt was not sufficient to even raise the question of jurisdiction of the judical machinery of the province of Quebec.

Also from Benediction Books ...
Wandering Between Two Worlds: Essays on Faith and Art
Anita Mathias
Benediction Books, 2007
152 pages
ISBN: 0955373700

Available from www.amazon.com, www.amazon.co.uk

In these wide-ranging lyrical essays, Anita Mathias writes, in lush, lovely prose, of her naughty Catholic childhood in Jamshedpur, India; her large, eccentric family in Mangalore, a sea-coast town converted by the Portuguese in the sixteenth century; her rebellion and atheism as a teenager in her Himalayan boarding school, run by German missionary nuns, St. Mary's Convent, Nainital; and her abrupt religious conversion after which she entered Mother Teresa's convent in Calcutta as a novice. Later rich, elegant essays explore the dualities of her life as a writer, mother, and Christian in the United States-- Domesticity and Art, Writing and Prayer, and the experience of being "an alien and stranger" as an immigrant in America, sensing the need for roots.

About the Author

Anita Mathias is the author of *Wandering Between Two Worlds: Essays on Faith and Art.* She has a B.A. and M.A. in English from Somerville College, Oxford University, and an M.A. in Creative Writing from the Ohio State University, USA. Anita won a National Endowment of the Arts fellowship in Creative Nonfiction in 1997. She lives in Oxford, England with her husband, Roy, and her daughters, Zoe and Irene.

Visit Anita at http://www.anitamathias.com, and on
http://theoxfordchristian.blogspot.com, her Christian blog;
http://wanderingbetweentwoworlds.blogspot.com/, her personal blog, and
http://thegoodbooksblog.blogspot.com, her literary and writing blog.

The Church That Had Too Much
Anita Mathias
Benediction Books, 2010
52 pages
ISBN: 9781849026567

Available from www.amazon.com, www.amazon.co.uk

The Church That Had Too Much was very well-intentioned. She wanted to love God, she wanted to love people, but she was both hampered by her muchness and the abundance of her possessions, and beset by ambition, power struggles and snobbery. Read about the surprising way The Church That Had Too Much began to resolve her problems in this deceptively simple and enchanting fable.

About the Author

Anita Mathias is the author of *Wandering Between Two Worlds: Essays on Faith and Art*. She has a B.A. and M.A. in English from Somerville College, Oxford University, and an M.A. in Creative Writing from the Ohio State University, USA. Anita won a National Endowment of the Arts fellowship in Creative Nonfiction in 1997. She lives in Oxford, England with her husband, Roy, and her daughters, Zoe and Irene.

Visit Anita at http://www.anitamathias.com, and on http://theoxfordchristian.blogspot.com, her Christian blog; http://wanderingbetweentwoworlds.blogspot.com/, her personal blog, and http://thegoodbooksblog.blogspot.com, her literary and writing blog.

CPSIA information can be obtained at www.ICGtesting.com
Printed in the USA
LVOW11*1813070516

487177LV00006B/98/P